Economics of Water Pollution

# Economics of Water Pollution
## The Indian Experience

M. N. MURTY, A. J. JAMES AND
SMITA MISRA



DELHI
OXFORD UNIVERSITY PRESS
CALCUTTA CHENNAI MUMBAI
1999

*Oxford University Press, Great Clarendon Street, Oxford* OX2 6DP

Oxford New York
Athens Auckland Bangkok Calcutta
Cape Town Chennai Dar es Salaam Delhi
Florence Hong Kong Istanbul Karachi
Kuala Lumpur Madrid Melbourne Mexico City
Mumbai Nairobi Paris Singapore
Taipei Tokyo Toronto

*and associates in*

Berlin Ibadan



ISBN 0 19 564395 X

Typeset by Excellent Laser Typesetters, Tri Nagar, Delhi 110 035
Printed in India at Rashtriya Printers, New Delhi 110 032
and published by Manzar Khan, Oxford University Press
YMCA Library Building, Jai Singh Road, New Delhi 110 001

# Contents

# Tables

# Figures

# Appendices

# Preface

This book is meant to be a resource text for researchers, both students and professionals, who are interested in the topical and important area of pollution control. The book aims at presenting primary research material that takes the reader to the frontiers of applied environmental economic research in developing economies. It is the first of its kind in the Indian context.

The focus of the book is largely economic instruments and institutions used by large industries, by small-scale industries in industrial estates and isolated small-scale industries to combat industrial water pollution. The aim is to introduce readers to theoretical and applied approaches to pollution control in developing countries, so as to enable them to contribute further to this small but vital field of economic research.

The tenor of the chapters, thus, is to detail the analytical approaches and methodology used, with a view to providing a survey of state-of-the-art procedures for economic analysis in pollution economics, supported by empirical analysis of trends in the Indian context. It thus serves both as a documentation of empirical evidence on pollution control in the Indian economy as well as a technical resource book for students of environmental economics, who are interested in the analysis of pollution problems in a developing country context.

Most of the material presented here is based on a research project on industrial water pollution abatement in India undertaken by the authors during 1994–6, at the Institute of Economic Growth. Additional material has, however, been incorporated to generate a research text for applied environmental economics in developing countries.

This study has been undertaken with the financial support of the International Development Research Centre (IDRC), Canada. We are grateful to Dr Michael Loevinsohn, Dr Rohinton Medhora and Dr Aung Gyi of IDRC for their keen interest in this study since its inception. We are particularly thankful to Dr Loevinsohn without whose determination this study would have not been undertaken.

Many officials of the Central and State Pollution Control Boards have helped us by sparing time for discussions with our project team and organizing meetings of team members with the officials of various industries. We are grateful to all of them. We are particularly thankful to Dr Dilip Biswas, Chairman of the Central Pollution Control Board for promptly writing to State Pollution Control Boards asking them to cooperate with our project team. Some of our team members have also benefited from discussions with scientists from the National Environmental Engineering Research Institute (NEERI), Nagpur, and the Central Leather Research Institute (CLRI), Madras, during their visits to these institutes. We express our gratitude to them.

We have also benefited from the comments of participants in project workshops and seminars during the period 1994–6 and a national seminar on 'Environmental Regulation and Industrial Water Pollution Abatement in India' during 21–2 October 1997 at the Institute of Economic Growth, Delhi, and in seminar presentations at the Delhi School of Economics, Centre for Science and Environment, Delhi, the University of Hyderabad, the Indira Gandhi Institute of Development Research, Bombay, and the Centre for Development Studies, Trivandrum. We express our sincere thanks to all of them.

We are grateful to Dr U. R. Prasad, a Research Associate with the project since its inception, for taking enormous interest in the project work. He is a co-author of two chapters of this book and has participated along with the project team in data collection, processing, and analysis. We are also indebted to a number of enumerators from different parts of the country who helped us collect data.

We are grateful to Professors S. N. Misra snd S. R. Hasim, the former Directors of the Institute of Economic Growth (IEG), and to Professor Pravin Visaria, current Director, for providing us a lot of encouragement and support in executing the project work. Also, Mr T. C. Sharma and other staff members of the IEG administration, and Mr K. Lal and staff at IEG Computer Centre have provided us with full cooperation without which we could not have carried out this work. We express our thanks to them.

We also thank Ecotech Services (India) for providing computer and printing facilities to prepare the final draft of this book.

We also wish to mention our debt to Ms Nitasha Devasar, senior commissioning editor, and Debolina Baruah, editor, of the

Oxford University Press, New Delhi, without whose constant encouragement and support this book would not have reached the stage of publication.

M. N. Murty
A. J. James
Smita Misra

New Delhi
April 1998

# General Introduction

## 1. INDUSTRIAL POLLUTION: A THREAT TO ENVIRONMENT

Industrial pollution has been and continues to be a major factor causing the degradation of the environment around us, affecting the water we use, the air we breathe and the soil we live on. But of these, pollution of water is arguably the most serious threat to current human welfare. Water is polluted not only by industries but also by households. Both industries and household waste water contain chemicals and biological matter that impose high demands on the oxygen present in water. Polluted water thus contains low levels of dissolved oxygen as a result of the heavy biological oxygen demand (BOD) and chemical oxygen demand (COD) placed by industrial and household waste materials discharged into water bodies and water systems, both above and below the earth's surface. In addition to low levels of dissolved oxygen in water, industrial wastes (effluents) also contain chemicals and metals which are directly harmful to human and ecosystem health. Arsenic, lead, mercury, cadmium and zinc are examples of (heavy and other) metals that are found in water bodies and water ways polluted by industrial discharges. The supply of water through river valley projects and ground water extraction thus has repercussions for the health and safety of people.

Apart from health effects, which indirectly affect human (read labour) productivity, polluted water also affects land productivity. Crop production suffers from using contaminated irrigation water from both surface sources and from groundwater aquifers. There is now growing evidence from many developed and developing countries on water-related effects on land productivity, and human health in the command areas of major river valley projects. River valley projects also affect the ecology of the valley with serious consequences for biodiversity. A related environmental impact of human activity on water quality is the increasing evidence from coastal areas of contamination of fresh groundwater aquifers with sea water as a result of excessive withdrawal of groundwater.

Air is polluted largely by atmospheric discharges from fossil fuel combustion and from toxic chemicals from industrial production. Not only do emissions of gases like carbon dioxide, carbon monoxide, sulphur dioxide and nitrous oxides affect local populations, but they can also affect the global environment. These gases, some of which are termed greenhouse gases (because of their contribution to the greenhouse effect), and others like chlorofluorocarbons (CFCs) are responsible for stratospheric ozone depletion, and for phenomena like global warming which can trigger global climatic changes. The control of air pollution reduces damages from air-borne pollutants, with benefits ranging from local improvement in human, animal and plant health to global benefits as a result of controlling phenomena like global warming and ozone depletion.

Fortunately for the human race, some environmental resources are regenerative. The capacity of the environment (especially air and water) to assimilate pollution loads up to certain levels have led to these environmental characteristics being termed 'life-support systems'. But there are limits to these capabilities, which we would be wise to preserve. For, increasing pollutant loads beyond the 'assimilative capacities' of environmental receiving media like air and water can lead to an erosion of these capacities. The control of industrial pollution will go a long way towards preserving these special environmental resources at their natural regenerative levels. The benefits from the preservation of environmental resources are manifold. Not only can water pollution abatement provide marketable benefits such as reduced damage from water borne diseases, savings in the cost of supplying water for household, industrial and agricultural uses, control of land degradation and fisheries development, it can also generate non-marketable benefits like improved environmental amenities, aquatic life and biodiversity. There is however the growing fear that it may even now be too late to reverse many trends already set in motion. That should be sufficient reason to accord high priority to the control of environmental pollution so as to minimize future impact and to reverse those effects which are still within our grasp.

The need for such action implies that whatever corrective measures are currently being taken are somehow not enough, or just plain wrong. Strangely enough, economic analysis of environmental problems points out that the market is largely to blame—either by sending out the wrong signals, or by not being there at all. And one major reason for this state of affairs is the lack of information on the quantity of costs and benefits associated with the consequences of both, the problems and their solutions.

## 2. A CLASSIFICATION OF ENVIRONMENTAL PROBLEMS

Environmental problems can be classified in many ways and so can the policy instruments proposed and implemented for their solution. One basic division is into resource depletion problems and pollution problems. The former refer to problems with the use of renewable resources such as forests, fisheries or biodiversity, and to problems associated with the draught on non-renewable resources such as coal, oil and minerals. The latter deal with the injection of wastes into the atmosphere from production or consumption activities. These can once again be divided either by source or sink[1]. The three broad categories arising from the latter classification are air pollution, water pollution and soil or land pollution.[2]

A second basic division is according to the magnitude of the environmental problem, in particular, whether or not they are

[1] 'Source' of the pollution refers to the pollution-generating agency or body, such as automobile exhaust (for air pollution), industrial pollution (for air, water or soil pollution) or loudspeakers (for noise pollution). The 'sink' of pollution refers to the pollution-absorbing agency or body, such as the atmosphere (which can be local, for example, over a city or a traffic junction, or global), water bodies (which can be lakes, ponds, streams, rivers or oceans), city dumps or solid waste disposal sites or just agricultural soil (which has to absorb agricultural pollutants).

[2] These distinctions, however, are not water-tight. For example, the depletion of the ozone layer in the atmosphere may be viewed either as a resource depletion problem (from the point of view of the quantitative reduction in the resource) or a problem of environmental pollution (from the point of view of the wastes that are emitted into the resource-sink).

contained within national boundaries. Environmental problems which spill across country borders either in their causes or their consequences may be termed global, while those which do not may be called 'local'. This classification may be represented by the following matrix, with examples comprising the elements of the matrix.

| | *Local* | *Global* |
|---|---|---|
| *Resource Depletion* | | |
| Renewable | Deforestation | Depletion of fish stocks in international waters |
| Non-renewable | Coal mining | Depletion of the ozone layer |
| *Environmental Pollution* | | |
| Air | Smog | Acid rain |
| Water | Ground- and surface water contamination | Industrial pollution of international waters & rivers flowing across countries |

Here, the focus is primarily on local water pollution problems. These can be various, depending on the agents responsible for the pollution, the nature of the sources and sinks of these pollutants and the type of pollutant being analyzed.

Concerning the first, a distinction can be drawn between problem of industrial pollution and of domestic sewage. These would tend to differ in terms of their chemical content and the values they yield for gross pollution parameters like biological-oxygen demand (BOD), chemical-oxygen demand (COD), dissolved oxygen (DO), suspended solids (SS), dissolved solids (DS), pH, etc.

An alternative classification is based on whether the pollution comes from individual 'point' sources or are indistinguishable on that score and are ascribed to 'non-point' sources. Typically, industrial effluents are 'point' sources since they can be identified uniquely with a given polluting unit. On a similar spatial scale, those industrial effluents originating from small-scale industries can be distinguished from those emanating from large-and medium-sized industrial units. Typical 'non-point' sources of water pollution include agricultural run-off, which contain fertilizer and pesticide residues.

Whether the pollutants are released into rivers, lakes or coastal waters can also be a source of distinction between types of water pollution problems. Also the siting of these industries, whether close to cities or villages, or in relatively isolated locations.

Another possible distinction is based on the type of output being produced by the polluting industry, e.g. tannery wastes versus distillery wastes.

And finally, the nature of water pollution can be differentiated on the basis of the dose-response relationships between the specific pollutants involved and their impact on human, animal or plant health of even general ecosystem functioning.[3]

Thus, a distinction can be drawn between problems caused by the presence of lead in the effluent wastewater discharged into a perennial river by a large industrial smelter located on the outskirts of a small township, and those caused by acidic discharges into a small lake by a small-scale pharmaceutical factory located in an industrial estate within a large city.

Each type of environmental problem, so classified, may call for a different policy approach, although it may be difficult to operationalize specific solutions to each problem. Nonetheless, it is useful to bear these distinctions in mind, both when attempting to analyze such problems and while reviewing the progress or failure of policies designed to tackle such problems.

The focus, henceforth, will be on the performance of policy instruments designed to counter environmental problems posed by the discharge of industrial wastewater into inland, or coastal water bodies.[4]

[3] Dose-response relationship basically refers to the impact a particular dosage of a pollutant (toxin) has on an organism (animal, plant or human) or ecosystem. Toxicology notes certain 'threshold' levels of dosage below which the organism (or ecosystem) may be tolerant to the pollutant (i.e. may not show ill effects), but beyond which the organism may show adverse health effects (chronic or acute). It is on the basis of such relationships that the Food and Agriculture Organization (FAO) and the World Health Organization (WHO) have defined Acceptable Daily Intakes (ADIs) of scores of chemicals (published as the *Codex Alimentarins*). However, these dose-response relationships are difficult to specify uniquely, since they vary according to 'internal' factors such as size, mass, age and sex of the organism, as well as 'external' factors, especially the presence or absence of other chemicals (Taub 1987).

[4] Though the problem of groundwater pollution is assuming alarming proportions, little has been done by way of explicit policy measures to tackle this important problem, and hence the problem of groundwater contamination is not discussed any further.

## 3. ENVIRONMENTAL PROBLEMS: EXTERNALITIES, PUBLIC GOODS AND PROPERTY RIGHTS

The definition of *pollution* in economics is based not only on the physical effect on the environment but crucially on the human response to the physical effect. The physical effect can be biological (e.g. change in biodiversity, or ill-health) or chemical (e.g. effect of harmful chemicals from industrial waste on aquatic life in a river). The human response is as a result of a loss of welfare (Seneca and Taussig, 1984, Pearce and Turner, 1990). Thus, while physical scientists base their perception of pollution solely on the physical effect on the environment, economists rely on perceived utility losses to recognize and define pollution. This allows pollution to be defined even in cases of activities where the physical or biological environment is not affected (e.g. noise pollution).

Environmental pollution is an 'externality' in welfare economics. An externality is present whenever individual A's utility and production relationships include real (i.e. non-monetary) variables, whose values are chosen by others (persons, corporations, governments) without particular attention to the effects on A's welfare (Baumol and Oates, 1988). An externality can be either beneficial (positive) or harmful (negative). Examples of beneficial externalities are, a neighbour's rose garden entering as a real variable in the utility functions of others living in the neighbourhood, and services of a light-house entering as a real variable in the production function of shipping companies. Negative externalities include noise pollution by aircrafts using an airport, which enters as a real variable in the utility functions of persons living in the neighbourhood, and the pollution of a river, which enters as a real variable in the production function of water-supply undertakings drawing water from the river.

Externalities possess characteristics of public goods (Bator, 1958). Public goods by definition are 'non-excludable' and 'indivisible' in contrast to private goods. A public good like defence services or street lighting is available for all (relevant) users unlike private goods like refrigerators or books, whose use is (usually) restricted to their owners. Also, a public good is undepletable in the sense that if one individual uses (more of) it, others are not deprived from using it. Environmental pollution, say, air pollution, is a public 'bad' since it affects simultaneously every resident in the area. It is an undepletable externality in the sense that one individual breathing polluted air does not alter the quality of air inhaled by another.

Externalities are also characterized by a non-convexity in the transformation function between two (or more) marketed goods, provided the commodity space includes these externalities. Thus, smoke from a power station can be a detrimental externality on a laundry, and if this negative externality is sufficiently strong, the technical transformation curve between the two marketed goods, electricity and laundry services will exhibit non-convexity (instead of the usually expected convexity).[5] Further, if the laundry quits the business with an increase in the intensity of the externality (smoke), further increases in the level of pollution will have no adverse effect, and there will hence be an inherent non-convexity (sometimes called a fundamental non-convexity (Starrett, 1972)) in the transformation function, which of course has to be defined to include this externality.

In welfare economics, externalities, public goods and non-convexities are seen as cases of market failure (the second welfare theorem). Arrow (1970) explains externalities and public goods as cases of missing markets. Arrow's thought experiment shows that in the case of public goods or externalities, voluntary trade is not possible and the cost of operating these markets is high (Cornes and Sandler, 1986). In this case, the failure of markets to evolve is also explained as a failure to establish and enforce property rights.[6] In the model of anarchy of Bush and Mayer (1974), property rights are a precondition for voluntary exchange—in other words, for the

[5] This example is from Baumol and Bradford (1972). The technical transformation function (TF) explains the rate of transformation of one commodity to another in a full employment economy. Convexity of TF implies that the opportunity cost (marginal cost) of producing X in terms of Y increases as more of X is produced in the economy. Non-convexity implies falling opportunity cost. Thus, in the presence of a strong externality due to smoke from electricity generation, the opportunity cost of producing electricity will be falling as more and more electricity is produced.

[6] The establishment and enforcement of property rights involves transaction costs. In the case of private goods, property rights are relatively easy to define and enforce, and hence markets exist for private goods. However, in the case of externalities and public goods, the cost of creating and enforcing property rights is prohibitively high, and hence markets for these goods are often absent or 'missing'.

evolution of markets. In many externality type situations, such as environmental problems, it is difficult to establish property rights since the exclusion of consumption is too costly to be privately profitable.

## 4. INSTRUMENTS OF ENVIRONMENTAL POLICY

One suggestion to deal with market failure arising out of environmental externalities is government intervention. The government has three basic choices: non-market policy instruments, market-based instruments and combinations of the two.[7] Non-market policy instruments include command-and-control (CAC) instruments, such as fines, penalties and threats of legal action for closure and imprisonment which, however, do not achieve the end result at minimum cost, and are hence not economically efficient.

Market-based instruments basically consist of pollution taxes (Pigou, 1932), marketable pollution permits (Dales, 1968), or hybrid combinations of the two, and are termed 'first-best' instruments because they guarantee market efficiency even in the presence of environmental externalities.

The choice between these policy instruments depends both on their efficacy in achieving the target level of emissions as well as on the relative size of welfare losses they produce (Baumol and Oates, 1988). There may however be cases where it is better to use a mixture of both taxes and permits, instead of either one of them (Roberts and Spence, 1976).

## 5. NON-GOVERNMENTAL ALTERNATIVES FOR POLLUTION REGULATION

Use of economic instruments for pollution regulation requires that the government or regulator be benevolent, which may not be the case in many developing countries as well as developed ones. Therefore, it is important to look for alternatives to government instruments. The alternatives can either be the collective action of all agents relevant for environmental regulation or purely a market option.

[7] Environmental policy instruments are discussed in greater detail in the Introduction to Part 1.

Coase (1960) provided an institutional alternative to governmental regulation for pollution control. He argued that many types of externalities can be controlled optimally by creating property rights and allowing costless bargaining between the relevant agents. In the Coasean solution, the government plays only a protectionist role by creating and protecting the property rights to an externality. It provides foundations for collective action for controlling pollution with the government playing a minimal role. There are now many examples from developed countries and from some developing countries showing the successful application of Coasean bargaining in dealing with environmental pollution.[8] Also, there is empirical evidence of the collective action of beneficiaries of local environmental resources resulting in a common property regime type of management (Chopra et al., 1990; Bromley, 1992; Murty, 1994a). The increased role of the market in the new institutions designed for environmental management is also evident from recent developments in the USA, Europe and other places leading to a dominant role for private agents in the formulation and enforcement of environmental regulation.[9]

## 6. VALUATION AND ACCOUNTING OF NATURAL RESOURCES

Valuation of damages from pollution or the benefits from pollution control is important for various reasons. It is needed, first of all, to justify investments or defensive expenditures made in pollution abatement and, second, for developing a system of monetary accounts for environmental resources to estimate environmentally corrected GDP (EGDP).[10]

Environmental resources supply both private and public goods type of services and therefore the benefits from their preservation can be classified as marketable and non-marketable benefits.

[8] See Pargal and Wheeler (1996) and Murty and Prasad, and Misra and Murty in Part II of this book.

[9] See Introduction to Part III for details.

[10] Environmentally corrected GDP is the GDP that can be generated by keeping the quality of environmental resources at their natural regenerative levels. This can be estimated by developing the monetary accounts of benefits and costs of preservation of environmental resources at their regenerative levels and integrating them into a system of national income accounting.

Examples of marketed benefits are commercial timber, irrigation, fisheries, and water for industrial and household uses. Non-marketed benefits can be classified into user benefits, option value benefits (Weisbrod, 1964) and existence and bequest value benefits (Krutilla, 1967). User benefits are direct non-marketed benefits such as environmental amenity services which a user enjoys by visiting a site harbouring a preserved environmental resource. Option value is the amount an individual, who is unsure of whether he would visit a site (such as a national park), would be willing to pay over and above his expected consumer surplus to guarantee that the site is available to him if he wishes to visit it in future.[11] Individuals may place certain value (existence and bequest values) on environment regardless of the fact that they feel certain they will never demand *in situ* the services it provides, either because the mere existence of an environmental resource forms a significant part of their real income or, because they have a motive to bequeath the resource to future generation.

There are special methods designed to estimate the non-marketed benefits from preservation. Mitchell and Carson (1989) classify them as (a) methods based on observed behaviour and (b) hypothetical methods. The first set of methods includes travel cost, hedonic prices, and so on, while the second set of methods includes contingent valuation and its variants (Freeman, 1993).

Valuation of environmental resources is important to estimate environmentally sustainable income or EGDP. Estimation of EGDP requires the development of physical and monetary accounts for environmental resources and the establishment of linkages between them, and the assessment of environmental benefits and costs (UN, 1993a). Monetary accounts of environmental resources can then be integrated with conventional national accounts for estimating EGDP.

## 7. LIMITATIONS OF THE CURRENT STUDY

Perhaps the major limitation of the studies reported in this book is the ground they have covered. Industrial pollution can affect other environmental media besides water and air. Also, local problems

[11] Option value is the premium an individual places on a resource to avoid the uncertainty of its availability for his future use.

have been highlighted at the expense of global problems. However, comprehensiveness comes necessarily at the cost of detail, especially if the publication is to be a single volume.

A second limitation is that attention has been reserved for certain policy instruments, namely, economic instruments and alternative institutional instruments based on collective and non-governmental action. Command and control instruments, the major plank of environmental regulation till date, have not been dealt with in detail. The major reason for this is that it is the considered opinion of the authors that economic instruments for environmental regulation, when applied appropriately, have several advantages to offer over existing regulatory measures. The aim is to highlight these instruments, in the context of an empirical analysis, and to illustrate how they can be applied.

Another limitation is that, even within the range of economic policy instruments, the pros and cons of using marketable permit schemes to control industrial water pollution have not been analysed in sufficient detail, at either the theoretical or empirical levels. This is because of the paucity of work and experience in this instrument in developing countries.

Further, not all chapters report empirical work. Some solely discuss theoretical aspects of economic instruments and other institutions for environmental regulation.

There is a definite need for more empirical work on the implications for developing countries of having international agreements to use either carbon taxes or marketable permits instruments to deal with the problems of global warming and ozone depletion. This can be an important agenda for future research in India.

## 8. STRUCTURE OF THE BOOK

This book deals with some theoretical aspects of the environmental externalities discussed above. It relates to an important environmental externality, industrial pollution, and examines in particular, the pollution of water by industrial production. It is organized in three parts.

Part I discusses the design, use and impact of environmental policy instruments. The introduction to this part reviews and compares, albeit briefly, the various policy instruments available for government intervention. The first chapter examines the role of the government in dealing with environmental externalities, in the context of the Indian and the Japanese experiences, noting in particular, lessons India may learn from Japan. Chapter 2 is a review of environmental policy instruments in practice in both developed and developing countries which also notes the problems with regulatory systems in developing countries and certain important requirements for the effective use of economic instruments in these countries. Chapter 3 deals with the practical problems of designing a pollution tax for industrial water pollution abatement in India, specifies a methodology to do so, and provides estimates of pollution tax rates for industrial water pollution control using a 'taxes-and-standards' approach.

The chapters in Part II highlight certain theoretical and empirical aspects of collective action as an alternative institution to deal with industrial water pollution. Chapter 4 explains collective action as an institutional alternative to economic instruments. Chapter 5 examines the role of the community, and of location-and firm-specific factors in influencing the compliance of industries with environmental standards, and establishes that local, community-specific factors like the level of economic development and political participation of communities play a significant role in industrial compliance. Chapter 6 examines, with case studies, the possibilities of collective action to deal with water pollution in industrial estates containing small-scale industries in India. Chapter 7 presents a model of collective action which serves as a deterrent to a non-benevolent government.

Part III begins with the analysis, in Chapter 8, of an important problem in Natural Resource and Environmental Economics, the problem of natural resource accounting, in the context of estimating environmentally adjusted Gross Domestic Product (EGDP) in the manufacturing sector of India. Chapters 9 and 10 are based on a detailed case study of an industrial estate in India, and provide estimates of benefits and costs of industrial water pollution abatement. This case study is a pioneering attempt in using contingent valuation methods to estimate benefits from industrial water pollution abatement. Apart from discussing the potential and problems of using contingent valuation methods in developing countries such as India, it provides a practical methodological illustration of the estimation of benefits from water pollution.

The concluding chapter draws together the main points from the rest of the book, and also highlights the research agenda for the future in order to assist and support the evolution of economic instruments for the effective regulation of industrial pollution in India.

# PART I
## Policy Instruments for Pollution Control

# Introduction to Part I

## ENVIRONMENTAL REGULATIONS IN PRACTICE

Developed countries and many developing countries have comprehensive legislation for environmental regulation. This legislation provides for the use of command and control measures, economic instruments and institutions facilitating peoples' participation for environmental management. However, developing countries including India lag far behind in implementing them. In many of these countries, command-and-control measures are more commonly used. Many developed countries however are now using economic instruments—pollution taxes in West European countries and marketable pollution permits in the USA—in combination with command-and-control measures. Japan has decentralized regulation in which local governments, communities and industry play an important role.

The role of government in environmental regulation may be coercive or catalytic, depending upon the instruments used and the institutions chosen. Command-and-control measures are regarded as being coercive and may result in an inefficient management of the environment, especially if the implementation of environmental legislations is centralized and the government is non-benevolent. Government has to play a big role in the design and implementation of economic instruments and therefore these instruments become ineffective with a non-benevolent government. In particular, implementation of legislation empowering local communities to deal with environmental problems requires the government to play an enabling role. It has to facilitate peoples' participation by providing a strong legal framework and financial incentives to local communities.

In actual practice, both developed and developing countries use a combination of instruments and institutions for environmental regulation.

## 1. POLICY INSTRUMENTS FOR POLLUTION CONTROL

Policy instruments can broadly be classified into three: (1) command-and-control or regulatory instruments (or simply direct regulation), (2) economic instruments or incentives, and (3) suasive instruments.[1] However, these are not necessarily mutually exclusive categories, and an environmental policy to reduce industrial waste water discharges may incorporate two or more of these options. There are a variety of instruments within each group, and these are discussed in greater detail below.

### 1.1 Command and Control (CAC) Instruments

Such policy instruments are again of two broad types; *technology-based standards*, that define specific abatement techniques for each source of pollution; and *performance standards*, which set specific ceilings on emissions for each source, leaving the source to determine the most effective way to achieve them (Oates, 1985). The second thus refers to restrictions or prohibitions on the emissions of certain pollutants, where punishment for excess is administered through the legal system and can range from imprisonment and fines to the forced closure of the offending industrial units. A variation of the latter forces all industrial units to abate by a uniform amount, irrespective of the efficiency characteristics of these industrial units.

There are theoretical and practical problems with the use of CAC instruments. Theoretically, it can be shown that the costs of compliance are generally higher when CAC instruments are used than when economic incentives (like taxes or marketable permits) are used (Baumol and Oates, 1988). Under a pure CAC regime, there may also be little incentive for firms to innovate or invest in more efficient pollution control technologies or in cleaner process technologies. In practice, however, CAC instruments may ensure a more certain level of pollution abatement, and may also be able to effect rapid improvements in ambient environmental quality. Actual experience, however, would depend not only on the mix of policy instruments within the environmental policy enforced, but also on the commitment and enforcement capability of the implementing government (see section 2.2.1 in Chapter 2 below).

[1] The term 'suasive instruments' is from O'Connor, 1996.

## 1.2 Economic Instruments

Such policy instruments, by contrast, seek to provide economic incentives for various industrial units to abate the level of emissions from their production activities, in a variety of ways. Economic instruments can be divided into three basic categories:

- price-based instruments;
- quantity-based instruments; and
- hybrid instruments.

Price instruments are basically variations on the theme of taxing pollution-causing agents for the negative externality (and therefore social costs) they impose on the population at large. The basic analytical approach to environmental taxation was provided by A. C. Pigou in his seminal work entitled *The Economics of Welfare* (first published in 1920). Given the centrality of his contribution, it is probably worthwhile going over the basics of this approach before discussing quantity controls and suasive instruments.

### *1.2.1 Price-based Instruments: The Pigouvian Approach*

Pigou first suggested using a combination of taxes and subsidies to deal correspondingly with detrimental and beneficial environmental externalities. In the case of negative externalities, Pigou advocated what has come to be known as the Pigouvian tax on output as a corrective economic instrument to realize socially optimal levels of economic activity. This is illustrated in Fig. 1, where MSC stands for marginal social cost, MSB refer to marginal social benefit, $O^*$ is the optimal level of output and TAX stands for the optimal tax rate.

Equating the marginal social cost of pollution (which rises with increases in pollution-causing output) to the marginal benefits achieved from the production (which falls with rising output) produces the socially optimal level of output (O*) and the optimal tax rate (TAX).

Pigou's output-based analysis, however, requires the determination of the pollution coefficients of each level of output (i.e. how many units of pollution are caused by a single unit of output). Since this can be problematic, this analysis was translated into a tax on the effluents generated during production. The procedure is similar: by equating the marginal costs of effluent control (abatement cost)

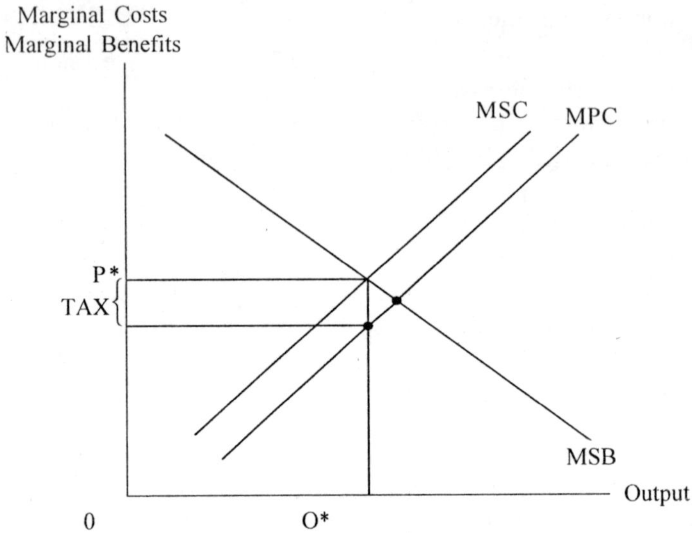


Fig. 1

to the marginal cost of damage (caused by the effluent being discharged into the environment), the socially optimal level of pollution (effluent) and the optimal tax rate are determined simultaneously. This is illustrated in Fig. 2 below, where MCA is the marginal cost of abatement, MDC is the marginal damage cost schedule, P is the socially optimal level of pollution and T is the optimal tax rate required to achieve P.

The models of general economic equilibrium with externalities that followed Pigou's original analysis (Baumol, 1972; Baumol and Oates, 1988; Murty and Nayak, 1982; Cornes and Sandlers, 1986) have shown that with Pigouvian taxes and subsidies, the competitive equilibrium of an economy with environmental externalities is Pareto optimal. The Pareto optimality with externalities requires that the marginal environmental pollution abatement cost of the generator is equal to marginal damages of the receivers in addition to other optimality conditions for the supply and demand for private goods. It has been shown that given market equilibrium prices for private goods and inputs, a tax (subsidy) on the generator of a detrimental (beneficial) externality that is equal to marginal damages (benefits) to the receivers at the optimum can make the competitive market

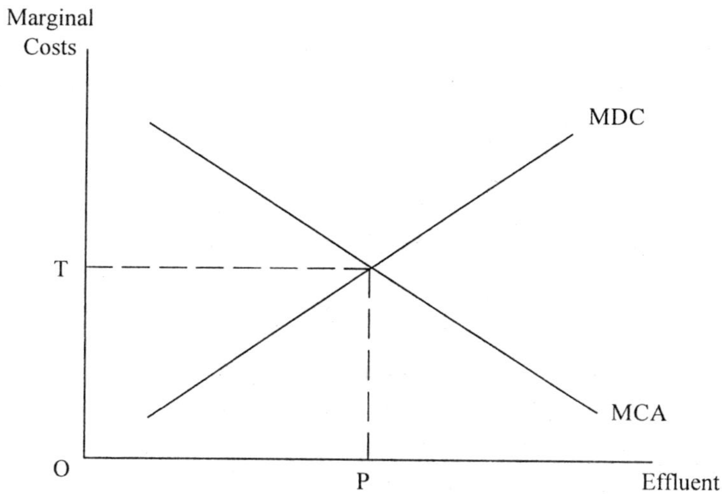


Fig. 2

equilibrium Pareto optimal.[2] It is in this sense that the Pigouvian (pollution) tax is often referred to as a 'first-best' instrument.

Three major problems, however, characterize the Pigouvian approach. First, the marginal damage cost schedule may be difficult to ascertain comprehensively and uniquely. Consequently, the optimal level of pollution may not be the level actually achieved by the tax. Second, the tax rate will have to be periodically revised upwards or downwards depending on whether the actual level of pollution is below or above the optimal level (see *inter alia*, Baumol and Oates, 1988, for a more detailed exposition). Third, the effectiveness of a pollution tax is constantly undermined by inflation and thus an effective tax (assuming it is pegged at the correct level initially) would have to be indexed to inflation, to ensure continuing effectiveness.

Since the Pigouvian prescription, economic (price-based) instruments for environmental pollution have been refined and redefined in various ways. These are briefly described below.

[2] Pareto-optimality requires that, in addition to the usual optimality conditions for private goods, the marginal environmental pollution abatement cost to the polluters (of abating pollution) is equal to the marginal damages of the victims.

*Emission Charges:* A (Pigouvian) tax on emissions specifies a rate of tax per unit of effluent discharged by the industrial unit into a water body. These charges may be uniform for all industrial discharges of a particular pollutant, or may vary across industries, or may even be specific to certain localities.

*Proxy charges:* Where monitoring of effluents is not feasible, either technologically or economically, a tax may be levied on a proxy commodity. This proxy may be the output of the firm, in which case the tax is an output charge (similar to the original Pigouvian tax), or it may be on an input used in production, whose use is proportional to the amount of pollutant produced in the effluent. In the latter case, the tax is termed an input charge.

All taxes increase the per unit costs of production and are designed to make the waste-producer face the true social costs of production. Nevertheless, the difficulties mentioned earlier of measuring the social costs of production and of designing a socially optimal tax rate indexed to inflation has meant that, more often than not, taxes have only provided a source of revenue to governments.

But revenue to the government is a loss of income for tax payers, producers and consumers. An alternative is for the government to subsidize pollution abatement in industry, and there are a variety of ways in which the government can do this including grants, loans or tax and duty allowances.

*Grants and Loans for Pollution Abatement:* Industries can be given grants or low-interest loans to purchase or install pollution control equipment, for research and development in pollution abatement, or to train personnel in effluent control techniques.

*Tax or Duty Concessions:* The government can reduce or abolish taxes on investment in pollution control equipment, or provide a concession on the customs duties charged for the import of pollution control technology.

*Depreciation Allowances:* The government can also stipulate a higher level of depreciation on pollution control equipment, enabling firms to reap accounting benefits.

Subsidies, however, can have detrimental effects on environmental quality (e.g. subsidies on fertilizers) and can also promote costly and unsound development (O'Connor, 1993). Thus, notwithstanding its theoretical standing as a first-best policy instrument, these problems and the uncertainty about the environmental improvement that pollution taxes can bring about prompted the search for more effective ways to make polluters pay.

### 1.2.2 Quantity-based Instruments

*Marketable Permits:* One alternative is to specify the optimal level of pollution, and accordingly, a 'standard' for environment (e.g. air) quality, and then try to achieve it in a cost-effective manner. An instrument that achieves a given standard at the lowest cost to society is the marketable permit system. Used primarily in the context of air pollution, this approach first determines the 'optimal' level of total pollution in a given geographical area. This level of total pollution is then divided into a number of permits among various polluting units within the area (either by free distribution or by auction). Firms which are comparatively more efficient in controlling their wastes (i.e., face lower unit costs for pollution control) may not only be able to continue their original level of production and emissions but may even have some pollution permits to spare. These surplus permits may then be sold to firms which are less efficient in controlling their wastes, i.e., face higher unit costs for pollution abatement. Provided that monitoring is possible and effective, the net result is that total pollution is kept within the prescribed (predetermined) levels. More efficient individual units will sell their surplus permits to less-efficient units which require more permits in order to continue with their original level of production.

This approach was first suggested by Dales (1968) by proposing the creation of property rights for environmental resources and then offering their sale to the highest bidder in the market. Quantity instruments like marketable pollution permits are also first-best instruments to control environmental pollution, in the sense that they make competitive equilibria Pareto optimal even with environmental pollution. Such quantity instruments can hence be alternatives to price or tax/subsidy instruments. Given the Pareto optimum quantity of emissions, a regulatory authority leaves the allocation of these emissions among sources to market forces. The regulatory authority issues permits for waste discharges such that, in total, discharges would be at the level that equates marginal abatement cost to marginal damage. Trading of permits among sources then establishes the market clearing price.

A major problem with this system, however, is that it requires geographical concentration of the polluting firms. In the case of air pollution, for instance, the total pollution has to be defined with respect to a geographical area. All air-polluting firms within this 'bubble' can then be included within the permit system. Geographically dispersed firms will make the definition of this 'bubble' problematic.

Another problem with the scheme is enforcement of the permitted levels of pollution. In the case of water pollution, for instance, even if firms in and around a lake (or a part of a river) are covered by a transferable permit scheme, without a strict check on whether or not these firms are actually discharging only the permitted level of pollutants, the scheme will not work. This will require a constant monitoring of the quantity and quality of wastes discharged into the water body, and not merely of the quality of wastewater generated or treated by the firm.

The difficulties of these 'pure' economic instruments can be addressed by combining the desirable features of each into a hybrid instrument. Indeed these can be combined also with a CAC framework. However only combinations of economic instruments are considered here.

### 1.2.3 Hybrid Instruments

*Taxes-and-Standards:* An economic instrument that combines the good features of both the above approaches is the taxes-and-standards approach, which seeks to tax emissions of wastes above the standards prescribed by the regulating authority. As in the marketable permits approach, a predetermined standard of environmental quality is set in place but emissions over these standards are taxed. This tax however is at a rate higher than the marginal cost of effluent treatment. In a pure tax system, the polluting firm will find it economically meaningful to treat effluents only upto the level where the marginal cost of abatement or treatment is equal to the tax rate. At higher levels of pollution abatement, it will find it economical to pay the (lower level of) tax rather than incur the higher costs of treating the effluent. If the tax rate is set at a higher level than the cost of treatment, the firm will find it economical to treat the pollutant rather than to pay the tax. (see section 3.2 in chapter 3 for further details).

*Taxes-and-Rebates or Two-part Tariffs:* One way to provide industry with some incentive to persist with their investment in pollution control technology is to give them some reward for reaching the standards laid down for environmental quality. In the case of a tax levied on every unit of effluent discharged by the firm, this reward can be in terms of a lower rate of tax on the units of effluent discharged that comply with the standards. For instance, if a firm has a total of 110 units of effluent, out of which 80 units fall within the prescribed standards, then a reduced tax rate can be applied on the 80 units. Here, the reward for compliance—the reduced tax rate—may be viewed as a rebate paid by the government to industry.

*Taxes-Standards-and-Subsidies:* Another hybrid economic instrument is one where firms under a taxes-and-standards scheme are provided a subsidy on the marginal costs of abatement of pollution (e.g., Roberts and Spence, 1975). Typically, pollution control equipment is subsidized in order to encourage firms to comply with the standards of environmental quality.

Such an instrument works as follows: The regulator issues a number of emission permits creating a market for permits. Let the equilibrium price per permit that emerges in the market be p. At the same time, the regulator allows polluters to generate emissions in excess of quantity authorized by their permit holdings but charges a pollution tax, t per unit of such emissions. Finally, the regulator offers a subsidy, s per unit for any unused permits, where $s < t$. It can be shown that in equilibrium, the following result holds: $s < p < t$. This is true because if p is greater than t, no one would purchase permits but pay the tax instead so that p would have to fall. On the other hand, if s exceeds p, it would be profitable to purchase as many permits as possible and hold them unused at a profit of (s–p) per unit; but obviously nobody would be willing to sell a permit at this price. The advantage of the mixed instrument is that it avoids too much or too little control of emissions than required (which may result from using either tax or permit instruments when there is uncertainty about the marginal abatement cost). By manipulating the values of p, t and s, one can realize optimal emission levels. For instance, if the regulator's estimated marginal abatement cost curve is higher (lower) than the true curve, there will be less (more) emission reduction than desired, under the permits instrument. If the tax t is lower than the true marginal abatement cost, it will pay the polluters with the mixed instrument to emit more than permits allow and pay the tax t on all emissions that exceed those justified by the permit holdings. Similarly, if the subsidy s is higher than true clean up costs, it will pay polluters to continue to reduce their pollution and hold excess licenses unused in return for the subsidy payment. Therefore, whenever a permit/tax instrument performs badly, the mixed instrument helps to transform it automatically into a tax/permit instrument.

The advantage of such a mixed instrument is that, given the uncertainty about the marginal abatement cost curve, it can adjust for the individual inadequacies of the two instruments (i.e. the greater-than-required or lower-than-required control of emissions).

### 1.2.4 Suasive Instruments

*Voluntary Pollution Agreements:* These are non-legally binding agreements, either between industry and the government or between industry and the local community, in which the industry in question 'volunteers' to reduce its waste emissions by a certain amount within a specified time. In some cases, however, the agreement is not purely voluntary, with the government exerting pressure on the firm in question to comply with the agreement reached, in the form of sanctions or hidden threats (for more details, see the Asian cases in section 2.1 of chapter 2).

The main attraction of such an agreement to the firm in question is the public approval that a successful abatement programme will bring. This again may be used by the government as a sort of negative pressure on defaulting firms or firms which cannot keep to their agreements—for instance, by publishing (or even threatening to publish) information pertaining to the results of (inadequate) abatement programmes undertaken by each firm.

Suasive instruments however are not the mainstay of environmental regulation, which continues to rely on direct regulation and economic instruments.

## 2. CHOOSING THE RIGHT INSTRUMENTS

The choice between the two first-best economic instruments depends both on their efficacy in achieving a target level of emissions as well as on the relative sizes of the welfare losses they produce. Baumol and Oates (1988) offer the following guidelines for this choice of (economic) policy instrument, based on the relative slopes of the marginal damage function (equivalent to the marginal benefits of abatement) and marginal abatement cost function (also known as the marginal control cost).

- If the welfare-maximizing regulator (the government) knows the damage functions and the abatement cost functions with certainty, it can impose either a pollution tax or a system of marketable permits, in order to achieve optimal control of the environmental externality. In other words, the final outcome and emission level obtained will be exactly the same, and the optimal tax to be set (under the pollution tax policy option) will equal the optimal permit price achieved (under the marketable permit system).
- This indifference between the choice of (first-best) instruments holds even if there is uncertainty about the marginal damage function (but certainty about the marginal abatement cost function).
- But if there is information uncertainty, especially about the abatement cost function, then the actual emission reductions obtained will vary with the choice of the policy instrument:
  - If the marginal abatement cost function is lower than expected, emission reductions are lower than required with a marketable permit system, and higher than required with a pollution tax.
  - And vice versa, if the pollution abatement cost is higher than expected.
- The welfare losses that result from the use of these two instruments varies depending on the relative slopes of the marginal damage and abatement cost functions. The special case under which these two welfare losses are exactly equal is characterized by the following:
  - the marginal abatement cost and damage functions are linear
  - the former is known with certainty, but the latter is uncertain
  - the slopes of the two curves are the same.
- If the marginal cost of abatement curve is steeper than the marginal benefits (of abatement) curve, the pollution tax is the preferable instrument, and vice versa.

Thus each instrument may be the better choice depending on the context. For instance, Harrison (1988) finds that in cases where marginal damages are fairly constant, as in case of airport noise pollution, the tax is preferrable to a system of permits. Further, in cases of rapidly rising marginal abatement costs, as is the case with the control of automotive exhaust fumes, quantity instruments are preferable to tax instruments.

It is important from the point of view of designing environmental policy to examine factors favourable for the choice of one or the other instruments discussed above. Taking up the case for choosing pollution tax against other instruments first, one can put forth various arguments for selecting the pollution tax. One argument (which figures prominently in the current debate) is that a pollution tax has a 'double dividend' property i.e., while controlling environmental externality, it also earns revenue to government which can be used to decrease other distortionary taxes. (Oates, 1991; Pearce, 1991; Rajah and Smith, 1993). This is the weaker version of the double dividend hypothesis, and the stronger version argues that the deadweight losses will be higher than the loss of environmental benefits due to a pollution tax or marketable permits.[3] But while the weaker version of double dividend hypothesis may hold, the stronger version may not (Boovenberg and Mooij, 1995; Knoester and Sindern, 1995).[4] Another argument in favour of pollution tax is with reference to equity. Since environmental resources belong to society at large, those who use them must compensate the owners. The pollution tax does just this because it is based on the 'polluter pays' principle. Yet another argument in favour of pollution taxes is the transition cost of moving to a system of marketable permits. In the case of marketable permits, a series of transfers have to take place before the least cost allocation of permits materializes, resulting in significant search costs. However, in the case of a tax there is no need for such transfers; each source directly responds to the

[3] The loss in environmental benefits is in terms of benefits foregone from the developmental use of environmental resources due to use of pollution taxes.

[4] Environmental taxes typically exacerbate, rather than alleviate pre-existing tax distortions. Sandmo (1975) has shown that in the presence of distortionary taxes, the optimal pollution tax is lower than the Pigouvian tax. Intuitively, the collective good of environmental quality directly competes with other collective goods. Therefore, even if revenue from pollution tax is used to cut distortionary taxes, the marginal cost of environmental policy rises with the marginal cost of public funds. Using a general equilibrium model, Knoester and Sindern (1995) have shown that double dividend hypothesis is not confirmed. In a situation in which increased environmental taxes on polluting commodities are used for a reduction of direct taxes on labour, the higher environmental taxes are shifted forward into higher wages and prices. As a result, the positive effect of a simultaneous decrease in direct taxes on labour are wiped out. Hence the case for environmental taxes rests solely on environmental benefits.

incentives offered. Further, as mentioned earlier permits will not work if polluting units are geographically dispersed.

There are nevertheless various cases for the choice of marketable pollution permits over a pollution tax. First, permits can reduce uncertainty and adjustment costs in attaining the prescribed environmental quality. Given the uncertainty of polluters' initial responses to a tax and the possibility that the initial tax rate is not high enough to realize the standards, the tax may have to be raised again and again until the target level of emissions is reached. This can involve costly adjustments by polluters in the levels of waste discharge and abatement technologies. In contrast, in case of marketable permits the regulator directly sets the required emission levels which are then realized without further manipulation by the regulator. Secondly, with economic growth and inflation, the pollution tax has to be raised periodically to avoid the erosion of value of tax with inflation and to meet the growing demand for treatment of waste emissions. In the case of permits, market forces automatically translate the effects of growth and inflation into higher prices of permits. Thirdly, pollution taxes become cumbersome if they have to discriminate between polluters across geographical regions. Different rates of taxes have to be levied in different regions depending on the local ambient environmental quality (differences in water quality along different stretches of a river; locations with different levels of air pollution). A system of marketable pollution permits can deal with this spatial dimension of pollution in a less complicated way by allowing firms in different locations (but emitting wastes into a well-defined pollution sink) to trade at different prices.

There may however be cases where it is better to use a mixture of both taxes and permit instruments, instead of either one of them (e.g. Roberts and Spence, 1976).

The attractiveness of market-based instruments with their obvious advantage over the command and control instruments makes it difficult for economists and policy makers to abandon the idea of using them. The difficult problem of estimating damages from an environmental externality can be avoided by selecting a set of standards for an acceptable environment. The regulator can select a rate of tax so as to achieve the specific standards or he can issue a number of marketable permits that are consistent with the standards. It can be shown that these instruments which are known as tax-standards and permits-standards methods facilitate the cost minimizing choices of levels of externality generating activities (Baumol and Oates, 1988). The main criticism of these instruments is that the environmental standards may be set arbitrarily. The failure of the market to provide information about the optimum level of environmental quality may justify public action through the process of collective decision making for deciding acceptable environmental standards. However, standards fixed through such a process can serve as a rough approximation to optimal environmental quality. There is nonetheless a clear case for using these instruments in situations in which environmental externalities impose very high social costs and these costs can be significantly reduced through feasible decreases in the levels of externality generating activities.

There may however be situations in which command-and-control instruments are unavoidable. In several cases, the social cost of a particular activity depends on the variables beyond the control of those directly involved. For example, the effects of a discharge of effluents into a river depends upon the condition of the river at that particular point of time. Similarly, stagnant air can trap air pollutants, perhaps even collecting them until they become hazardous to life. Therefore, exogenous meteorological conditions may contribute to occasional crises requiring temporary emergency measures in the form of command and control regulation. Pollution tax rates cannot be changed on short notice to deal with emergencies; and even if the changes are effected, polluters' responses follow with a longer time lag. Marketable permits also result only in long term adjustments in environmental quality and are not suitable for emergencies. Command-and-control instruments on the other hand can be quickly operated to deal with the more than normal amount of emissions arising out of emergencies, since they do not require extra monitoring. Therefore, in practice neither economic instruments nor command-and-controls alone constitute an optimal regulatory strategy. The cost minimizing strategy to realize given environmental standards is a mixed strategy consisting of economic instruments and command and control regulation (Baumol and Oates, 1988).

While economic instruments can be used by national governments to deal with national or local externalities, alternative institutions are needed to support economic instruments to deal with transnational or global externalities. There is a need for international cooperation for using economic instruments to control global externalities like ozone depletion and global warming. International agreements to have uniform reduction of emissions by different countries (e.g., the Montreal Protocal to reduce emissions of CFCs), harmonization of domestic commodity taxes, and trade taxes are inefficient ways of controlling transnational pollution (Baumol and Oates, 1988; Hoel, 1991; Bohm, 1997). In contrast, agreements leading to the use of international emission taxes or tradeable emission permits are first-best policy choices. (Hoel, 1992; Hoel, 1977). However, in practice, there have to be international agreements for using both international emission taxes and domestic pollution taxes to deal with global externalities (Murty, 1996a).

# 1. Role of Government in Environmental Management[1]

M. N. MURTY

Historically, national governments have played roles varying from the minimal functions of policing and protection of its citizenry from external aggression, to all-pervasive functions of producing all goods and services in the economy, taking consumption decisions for its subjects, deciding about what people can speak and what they cannot, thus abrogating their economic and civil liberties. Modern governments have acquired more functions than required, by drawing strength from either market failure arguments or economic development requirements or both. The market failure arguments are related primarily to externalities, public goods, missing markets, imperfect information and increasing returns.

The arguments based on economic development have arisen mainly due to the need for government to deal with poverty and income distribution, to provide equal opportunity or rights to education, health, nutrition and housing to its citizenry, or the need to assume a paternalistic role by guaranteeing social security and limiting the use of certain drugs, and to act as a custodian of future generations by controlling global warming and environmental pollution, and conserving forests, biodiversity, etc.

Gradual increase in the role of the government over time has been reflected in the significant growth in the size of its budget during recent times. There are some interesting viewpoints in the literature about the growth of the size of the government budget or public sector in the developed economies. One of them portrays the public

[1] This chapter is drawn mainly from Murty (1995a; 1996c). I am grateful to participants in a workshop on 'International Economics and Economic Development', March 1996, at the Institute of Economic Growth, Delhi.

sector as a monolithic agent, a Leviathan, that systematically seeks to maximize its budget, irrespective of the desires of its citizenry. It is pointed out that decentralization of government or public sector is an effective mechanism to control this Leviathan's expansive tendencies which is possible with a federal form of government. The basic argument is that, just as competition in private sector exercises its disciplinary force, so also competition among different units of the government can break the monopolistic role of a large central government and restrain the growth of the public sector.

The government has to play an important role in the management of environmental resources as can be observed from the experiences of many countries including India. During the first half of more than forty years of development in India, the planning process was silent about the sustainable use of environmental resources. By the time the Government of India realized the need for environmental regulation as an integral part of its strategy for rapid industrialization, considerable damage had already been done to India's valuable natural resources like forests, fertile lands and water bodies. The second half of the planning period spanning two decades from 1970 to 1990, has seen a spate of legislations providing for environmental regulations in India. These legislations have empowered the central and state governments to create necessary institutions and instruments and to monitor their functioning. Inspite of more than twenty years of environmental legislation, India is lagging behind in the use of instruments like pollution taxes. Institutional or technological options, such as promoting collective action in the industrial sector, and the implementation of environmental laws for protecting public interest are not comparable to international standards.

However during the Sixth and Seventh Plan periods and the recently concluded Eighth Plan period, sincere attempts have been made to integrate the management of environmental resources with the national economic planning process. Resources are earmarked for environmental protection out of plan funds for many development projects, especially projects for energy, and other infrastructural development. Certain measures have been taken for environmental safety and the promotion of environment-friendly technologies in industrial production. Industries have been asked to prepare a detailed report identifying the sources of pollution by activity, and measures have been taken to control them. All projects above a certain size and located in environmentally fragile areas have been asked to obtain clearance from the Ministry of Environment and Forests. Environmental regulation law has finally come to India.

## 1.1 QUALITY OF GOVERNMENT

In the literature on Public Economics, the case for providing more responsibilities to the government due to reasons mentioned above has been made on the assumption that the government is benevolent. It is important to recognize that there are limits to the number of duties that can be assigned to the government even if it is benevolent. For example, if a benevolent government takes charge of all-pervasive duties, it may result in thwarting the growth of social and economic institutions that support the incentive structures necessary for raising living standards. The result is a stagnant economy with an all-pervasive benevolent government. Also, the efficacy of a benevolent government depends upon the instruments it uses to perform its assigned duties. For example, some of its duties from the point of view of economic development, like poverty and income distribution can be performed by using either market-based instruments like taxes and subsidies or through direct government production and distribution of goods and services. The latter option, while allowing greater government intervention in the market process, may destroy the incentive structure of market institutions.

An analysis of the levels and structure of expenditures of contemporary governments in the world economy brings out clearly the varying roles they play in developed as well as developing countries. The size of the government, expressed in terms of the percentage of either government expenditure or revenue in GDP, is larger in developed industrial countries in comparison to developing countries. The percentage of central government expenditure and revenue in GDP is 31.46 and 26.99 respectively in the developed countries while the corresponding figures for the developing countries are 25.40 and 21.10 (see Table A 1.1 in the Appendix to this chapter). In industrial countries, the social security and health claim around 47.32 per cent of total government expenditure while the corresponding figure for the developing countries is 18.81 per cent. In developing countries, the bulk of public expenditure is incurred for the supply of general goods and services (which can otherwise be efficiently supplied by the market), infrastructure and defence. In other words, production of private goods by the public sector results in avoidable losses in market efficiency in developing countries, while in industrial countries, public expenditures are incurred mainly for supplying merit goods, social security services and so on (which take care of temporal and inter-temporal equity problems), especially to deal with cases of market failure. Governments in both developing and developed countries are concerned about the inability of market forces to deal with the equity problem; but the instruments they use to deal with this problem are different. While many developed countries rely on taxes and subsidies, developing countries depend on public sector production of some private goods apart from relying on taxes and subsidies.

Governments of many developing countries and even of some developed countries are malevolent or non-benevolent, so that the responsibility of dealing with market failures and achieving some developmental objectives cannot be left to them. In the case of control of externalities like environmental pollution, the corrupt bureaucracy of developing countries often colludes with the external diseconomy-creating economic agents in augmenting the externality problem rather than controlling it. In such situations, there is no other option left to the parties affected by the externality but to take recourse to collective action or political influence for dealing with the bureaucracy and the perpetrator of the externality. Coase (1960) argues that there is no need for governmental instruments like taxes and subsidies to deal with the externality problem. Given the initial property rights, either to the externality-generator or the receivers, the costless bargaining between them in a free market will result in the optimal control of the externality, the final outcome being independent of the initial allocation of property rights. Becker (1983) has shown that the political influence exerted by pressure groups of people can have similar effects as having a benevolent government to deal with market failures. In his own words:

> The same analysis of competition among pressure groups, without the introduction of social welfare function or benevolent government, explains expenditures on defence and other public goods, taxes on pollution, and other government activities that raise efficiency, even when some groups are hurt by these activities (Becker, 1983).

However, bargaining cannot be costless as Coase originally proposed, and transaction costs are important in the world of Coase and Becker. Even with transaction costs, there can be significant net welfare gains from collective action, and the case for collective action will be reinforced if the government is malevolent.

Take the case of industrial water pollution abatement in India. More than twenty years of environmental legislation in India has not produced a pollution tax and the CAC methods used so far by the Indian government have not made a dent in industrial pollution control. This has also been the case with many other developing countries. One important reason for this is the absence of awareness among people about the extent of damage from pollution, and their inability to organize themselves for participating in the management of pollution abatement. In such a situation, industry and the non-benevolent bureaucracy have incentives to collude for violating the environmental standards. Industry can bribe the regulator to over-report its effluent quality or not report it at all.

There are several agents involved in the political economy of industrial pollution abatement: affected people, elected representatives, bureaucracy and industry. Incentives exist for the emergence of sub-coalitions of agents like, a coalition of affected people and elected representatives and another coalition of bureaucracy and industry. These coalitions can constitute politically active pressure groups. Murty (1995b) has shown that the competition among these pressure groups may result in the optimal control of pollution.

## 1.2 CATALYTIC ROLE OF GOVERNMENT

### 1.2.1 Management of Local Commons: A Case of Village Forests and Grazing Lands

Management of local commons like village forests and grazing lands provides a good example of the catalytic role of government (Wade, 1987; Chopra, Kadekodi and Murty, 1990 and Murty, 1994a) in economic development. Village commons can be regarded neither as pure public goods nor pure private goods.They belong to an intermediary category possessing some characteristics of public and private goods. The vast literature on externalities and public goods (Cornes and Sandler, 1986; and Cropper and Oates, 1992) shows that such resources are faced with the familiar free rider problem. The conventional solution to the free rider problem is government intervention with penalties and incentives to regulate the behaviour of individuals. The other option is voluntary collective action by a group of agents for providing a public good. In the case of management of commons, collective action might take the form of setting and observing a rule of restrained access to a common property resource. The public good might be the resulting sustainable use of the resource (Wade, 1987). Olson (1965) argues that voluntary collective action by individuals to achieve their group interest is not possible, and there is a need for coercion by an outside agency to achieve the group interest. In his own words, his main proposition is that:

> unless there is coercion or some other special device to make individuals act in their common interest, rational, self-interested individuals will not act to achieve their common or group interest.

In other words, in the case of management of commons, the free rider problem frustrates voluntary collective action and therefore there is a need for regulation by an outside agency. However, the experiences in both developed and developing countries suggest that management of natural resources like forests by the government is not successful. This failure is mainly attributable to prohibitively high costs of policing forest lands without people's participation.

There are limits to collective action, as there are limits to government regulation in the management of commons. However, the strong point in favour of the former is that it does not encroach on individual rights and freedom while the latter does. Voluntary collective action is possible if there is a complementarity between individual benefits and the benefits to the community, which can usually be the case with the preservation of commons. For example, if the productivity of private assets, like agricultural land and milch animals, depends upon environmental resources like forest lands which are sources of supply of water and fodder, the households owning these private assets may voluntarily contribute to the preservation of these resources. Since it is widely recognized that benefits from co-operation are significantly higher than the benefits from free market management of common property resources, it is important to look into the problems that threaten collective action. Some of the problems that threaten collective action are as follows (Murty, 1994a):

• Many rural households in the developing countries do not have adequate assets to harness increased resources from collective management of commons. The preserved grasslands and forest areas in a village economy will increase the supply of fodder and forest produce. Unless households own sufficient livestock, the improved supply of fodder cannot be converted into increased incomes.

• The investment by the community on irrigation dams and tanks is complementary to preservation for the irrigation potential of these structures depends upon the preservation of their catchment areas. Many village communities in the developing countries cannot mobilize resources for these investments.

• Also, the household asset distribution is skewed, such that only a few households can appropriate benefits from co-operation. For example, only land holding households get benefits from the community irrigation structures and therefore landless households have no incentive to preserve the catchment areas of these structures.

• High population growth rates in the communities managing common property resources will make benefits from preserved resources to each community member negligible, and so people have no incentive to voluntarily participate in preservation.

The degradation of village commons and forest lands that are predominantly observed in many developing countries may therefore be attributable to the presence of limits to collective action in the village communities. Therefore, the government can play an enabling role to promote collective action. For example, the government can give either a subsidy or a loan to village communities for creating community assets that are complementary to preservation. It can undertake land reforms to reduce skewness in the distribution of land in a village economy. The government can also provide alternative employment for populations in excess of threshold levels in the village communities. In all these cases, the government plays a catalytic role, and there is no coercion or infringement of individual rights in the appropriation of benefits. In a detailed study of five villages that are located in the lower Shivaliks of Haryana and Punjab states in India, Chopra, Kadekodi and Murty (1990) have found that the government or an outside agency has successfully played a catalytic role in the management of village commons and forest lands.

Murty (1994a) considers a model for an Indian village in which the government owns the nearby forest land, households have a stake in its preservation, since there is a complementarity between the productivity of their private assets like cattle and farm land and preservation of the forest land, and the government is prepared to share the incremental forest produce from preservation with the households. He has shown that the Nash non-cooperative equilibrium with profit-maximizing agents yields the Pareto optimal management of forest lands. In such a setting, household participation in the preservation is ensured if government plays a catalytic role in agreeing to share the benefits from preservation with them. However, if the government is not willing to share the incremental benefits from preservation with the households according to a mutually agreed principle, a household does not have the incentive to take into account the effect of its participation on the rest of the households in the Nash non-cooperative equilibrium. It however takes into account the effects of the rest of the households'contribution to preservation on its pay-off. In other words, each household takes into account the effects of preservation on the productivity of its private assets like cultivable land and milch cattle, but not the effects on the productivity of forest land which is owned by the government.This asymmetric behaviour of households makes the Nash non-cooperative equilibrium, without a catalytic role of the government, inefficient in the management of commons.

### 1.2.2 Management of Global Commons: The Design of an International Carbon Tax

The atmosphere is a global commons, the degradation of which affects all people on the globe. Global warming is a negative externality caused by greenhouse gas emissions. A solution to this problem requires collective decisions by all countries of the world about targets set for the emissions and the instruments used to realize them (Hoel, 1990; Schelling, 1991; and UN,1992). The international targets for the reduction of greenhouse gases may require the member countries to obtain much larger emission reductions than required to combat the problem of local negative externality. The instruments to realize the global targets are under the control of a global authority while the instruments to control the local externality are controlled by governments of individual countries. For example, in the case of carbon dioxide ($CO_2$ ) emissions, we may have a fiscal system in which there are national pollution taxes and an international $CO_2$ tax with different centres of control. In such an arrangement, there is a problem of sharing by member countries, of revenue collected through the international carbon tax. The problem of designing national and international carbon taxes is similar to the problem of public finance with many levels of government or fiscal federalism in the world economy. In such a setting, national governments design national carbon taxes while an international agency fixes an international carbon tax.

In any institutional set-up designed to obtain efficient reductions in carbon emissions in the world economy, an international agency or world government has to play a catalytic role to enable national governments to use abatement technologies. Hoel (1992) has shown that a uniform international carbon tax by an international agency combined with net transfers of revenue collected through such a tax to the national governments yields a Pareto efficient allocation of emissions among the countries. Murty (1995c) has considered an institutional alternative similar to the fiscal federalism of the world economy in which national governments design and levy national carbon taxes to raise revenue and to contain a local externality problem, and an international agency designs and levies country-specific or uniform international carbon tax to deal with the global warming or an externality problem, and transfers revenue raised to the member countries, depending upon the local atmospheric quality obtained by each country. In such a setting, it is shown that a solution to the Nash non-cooperative game in which the national government fixes a national carbon tax and other taxes to maximize the national welfare subject to a revenue constraint, and an international agency fixes international carbon taxes and transfers revenue to the national governments to maximize the world welfare yields second best allocation of carbon emissions among the member countries. In this case, an international agency plays only an enabling role, by levying the carbon tax and distributing the revenue to the member countries.

## 1.3 INDIAN EXPERIENCE

### 1.3.1 Laws Regarding Industrial Pollution Abatement in India

The Government of India has shown some concern about the degradation of environmental resources only since the early seventies, more than fifteen years after economic planning for industrialization had started in India. The enactment of environmental laws, like Wildlife Protection Act, 1972 and Water (Prevention and Control of Pollution) Act, 1974 has marked the beginning of an era of governmental regulation for the control of environmental pollution in India. Besides, the Government of India has undertaken a number of regulatory and promotional measures. The Central Pollution Control Board was constituted in 1989. A number of legislations have been enacted, including the Water (Prevention and Control of Pollution) Cess Act, 1977; the Forest (Conservation) Act, 1980; the Air (Prevention and Control of Pollution) Act, 1981; the Environment (Protection) Act, 1986; the Motor Vehicles Act, 1938, (amended in 1988), and the Public Liability Insurance Act, 1991.

Of the legislation mentioned above, the Water (Prevention and Control of Pollution) Act, 1974, amended in 1986; the Water (Prevention and Control of Pollution) Cess Act, 1977, amended in 1988; the Air (Prevention and Control of Pollution) Act, 1981, amended in 1988, and the Environmental Protection Act, 1986, are the most important laws pertaining to industrial pollution abatement in India. The Water Act has resulted in the creation of Central and State Pollution Control Boards with the aim of prevention, abatement and control of water pollution. The Pollution Control Board can demand information from any person or industry to guarantee the compliance with the Act. This information could include the amount of discharge of effluents into a stream or on land and details about the installation and operation of pollution control equipment. Non-compliance with the directions of the Boards is punishable by imprisonment of upto three months and a fine of upto Rs 10,000. An additional daily fine of Rs 5,000 can be imposed for continued non-compliance. If the non-compliance continues beyond a period of one year after the date of conviction, the offender is punishable with imprisonment for two to seven years with a fine. The Board, if necessary, can also close down certain polluting factories. Table A 1.2 of the Appendix to this chapter provides the list of industries declared as water-polluting industries by this Act.

The Air Act, 1981 with its amendments in 1988 empowers the Central and State Pollution Control Boards to prevent, control and abate air pollution. The principal functions of the Central Board are to advise the Central Government on matters pertaining to improvement of air quality and the prevention and control of air pollution, to coordinate the activities of State Boards, provide them with technical assistance, lay down national standards for air quality and collect and publish data related to air pollution. State Boards have similar functions at the state level. The state government on the advice of the State Board may declare any area as an air pollution control area, and prohibit the use of any fuel that may cause air pollution in that area. Also, the state government should ensure that the standards for emissions of air pollution from automobiles are laid down by the State Board. No person shall, without the consent of the State Board, construct any air-polluting industrial plant in the pollution control area. Failure to comply with the provisions of this Act by any individual can make him liable to imprisonment of upto three months or a fine upto Rs 10,000 or both. Continued non-compliance after conviction can make the individual liable to a fine of Rs 100 per day. Table A1.3 of the Appendix to this chapter provides the list of industries declared as air polluting industries by this Act.

The Water Cess Act, 1977 with its amendments in 1988, provides for levying an assessment on water consumed by certain industries and by local authorities. The main objective of this assessment is to increase the resources of Central and State Pollution Control Boards for the prevention and control of water pollution. The rate of assessment applicable to various water polluting activities vary from Rs 0.015 to Rs 0.075 per kilolitre (kl) of water consumed (see Table A 1.4 in the Appendix to this chapter). If some or all the activities liable to the payment of water cess install effluent treatment plants, such activities are entitled to a rebate of 70 per cent of the assessment payable. The proceeds of the cess will first be credited to the Consolidated Fund of India, and the Central Government may pay to the Central Board or State Boards after deducting the cost of collection. The Central Government has to take into account the assessment collected by the State Government in deciding the amount payable to the State Board. Wilful evasion of assessment by the concerned parties can make them liable to imprisonment of upto six months, or a fine of upto Rs 1,000, or both.

The Environment Protection Act, 1986 is intended to remedy the lacunae noticed in the earlier laws, and to serve as a single environmental legislation. This Act provides for the protection and improvement of environmental resources like water, land and air, as per the decisions taken at the United Nations Conference on the Human Environment held at Stockholm in June, 1972. This Act empowers the Central Government to take all such measures as it deems necessary, for the protection and improvement of the quality of the environment. In particular, the powers of Central Government include the co-ordination of actions taken by the State Governments, planning and execution of nationwide programmes for the pollution abatement, and laying down standards for emission of environmental pollutants from various sources.

### 1.3.2 Fiscal and Other Instruments for the Control of Pollution in India

Threats of closure of a polluting activity, and penalties and imprisonment for the offending parties as envisaged in various legislations described above are all of the CAC type of regulatory measures. Even though it is widely known CAC measures do not provide the necessary incentives to the polluters to choose the least-cost methods of pollution control (Cropper and Oates, 1992), the Government of India has so far resorted to such measures alone for controlling industrial pollution in India. On the other hand, fiscal instruments such as pollution taxes or marketable pollution permits, despite being coercive, provide incentives to industries for adopting the least-cost pollution abatement technologies. Ironically, no serious attempts have been made in India so far for using such instruments for industrial pollution abatement. The currently levied assessment on the consumption of water by industrial activities cannot be treated as a pollution tax, since, as already mentioned, its main objective is to raise revenue for the Pollution Control Boards. As such, the assessment collected is very nominal (Rs 0.015 to 0.075 per kl) which normally does not have much effect on the demand for water by industries. Some of the recent research studies on the water pollution abatement in India (Dasgupta and Murty, 1985; Gupta, Murty and Pandey, 1989; Mehta, Mundle and Sankar, 1993) have found that the pollution tax on industrial water use should be several times higher than the current rate of water cess if the prescribed water quality standards are to be realized. One study carried out in the year 1989 (Gupta, Murty and Pandey, 1989) estimates the cost of treatment per kilolitre of residual water at 1987–9 prices at Rs 3.60 for paper and pulp industry, Rs 2.61 for oil refineries, Rs 2.21 for chemicals and Rs 1.64 for sugar. Another study carried out in 1993 (Mehta, Mundle and Shankar, 1993) estimates the marginal cost of abatement for the reduction of 100 milligrams of BOD in the residual water of the paper and pulp industry at Rs 0.38 at 1991–2 prices.

Another anomaly in the choice of instruments for the control of industrial pollution in India is the widespread use of subsidies for the polluters. It is well known that pollution taxes should be imposed on a polluting activity to make the necessary correction for the market failure. On the other hand, pollution subsidies provide perverse incentives to industries for increasing their scale of production, and hence environmental pollution (Baumol and Oates, 1988). The Government of India has been providing pollution subsidies to the industries in various forms. Pollution control equipment has been given the benefit of accelerated depreciation for corporate tax computation. In the 1993–4 budget, 100 per cent of the written down capital cost of such equipment had been exempted from the corporate tax. There has also been an investment allowance for corporate tax calculation for investments in pollution control equipment. There is another direct tax concession for the pollution control in terms of exemption of capital gains tax, for the capital gains obtained through the relocation of a polluting factory. There are also rebates of excise and customs duties for many types of pollution control equipment. These tax concessions or subsidies are not justified either on efficiency or equity grounds. They do not guarantee efficiency because they do not provide the polluters with incentives to choose the least-cost pollution abatement technologies. They also have the effect of increasing the polluters' profits, which is not desirable in a developing country, since the big factory owners/polluters normally belong to higher income groups.

There is one important case in which subsidies or financial incentives for the industrial units are justified for the control of industrial pollution in India. During the early sixties, several state governments in India came out with legislation in aid of industrial estates with the objectives of promoting small-scale industries and achieving balanced regional development. More than 800 items were reserved for exclusive production by small-scale enterprises. The small-scale enterprises are also entitled to financial assistance, tax benefits and subsidized electricity and water charges. As a result of this policy, the industrial estates in various parts of the country are dominated by small and medium-scale industries creating a number of environmental problems. Some of these problems are, inadequate understanding of the technology of waste generation and treatment, lack of required space for pollution control facilities, ineffective supervision and management of even simple installations for pollution control and the lack of technical assistance. Therefore, governmental assistance to these enterprises may be justified for helping them to circumvent these problems.

## 1.4 THE JAPANESE EXPERIENCE

### 1.4.1 Structure and Level of Government

The Japanese government consists of national government and local governments. The latter consist of prefectures and city governments. The national government levies personal income tax, corporate income tax and inheritance tax and indirect taxes like liquor tax, gasoline tax, and other commodity taxes. The major local taxes are the local income tax, property tax, and business tax. The total revenue of the central government constituted around 15.62 per cent of GDP during the fiscal year 1991. Even though the Central government may be collecting almost double the amount of revenue collected by the local governments, it transfers a sizeable part, almost 40 per cent, of it to the local governments.[2]

Given that the governmental intervention in the market process is minimal, Japan's economic success is mainly attributable to the system of industrial management which favours growth strategies and innovation, and encourages intra-firm and inter-firm competition. What has helped Japan has not been governmental policy as such, but the speed at which the Japanese government reduced its involvement in industries, in order to make better use of the market mechanism. Ironically, then, Japanese success may have come, not from an increase, but from a decrease in governmental intervention (Odagiri, 1994).

A look at the Japanese industrial policy during the period starting immediately after the war until now, clearly indicates that the rapidly declining role of government has been accompanied by economic prosperity. The history of Japanese industrial policy may be divided into five periods. The first period, covering upto 1950 after the war, is a period of priority production with supportive economic instruments like subsidies and loans, and allocation of input quotas to some priority industries like steel and coal. The second period, covering the fifties, is the period of industrial rationalization, with the governmental policies for protecting new and infant industries through tariffs and quotas. The sixties constitute the third period, the period of the new industrial order, with the liberalization of trade and foreign exchange. The Ministry of International Trade and Industry (MITI) has played a catalytic role by providing administrative

[2] See Matsuoka and Rose (1994) for details about Japanese Economic Statistics.

guidance to industry, with requests, suggestions and warnings. During this period, MITI has encouraged mergers and investment coordination among small industries to reap scale economies for competing with industries in other industrial nations.

The fourth period corresponds to the period after the oil crisis of 1973. During this period, the government, that is MITI, has encouraged knowledge-intensive industries, such as high technology and information processing industries, through administrative guidance. This period has also seen the Law for the Stabilization of Designated Depressed Industries and legalization of formation of cartels. Finally, the fifth period covering the eighties upto now may be regarded as a period of trade surpluses. MITI's main task has been to encourage imports and discourage exports, for instance, by persuading industries to adopt voluntary export restraint.

### 1.4.2 Environmental Policy

Environmental policy in Japan, like its other socio-economic policies, is based on the deeply-held values of consensus and cooperation. Wallace (1995) provides a comprehensive account of environmental policy in Japan. The policy has been based on a consensus approach involving local communities, bureaucracy and industry. In Japan, the local government bodies, like prefectures and metropolitan authorities, enjoy considerable autonomy (Local Government Law, 1947) and play a major role in balancing the interests of local communities and those of industry, in the context of environmental management.

Japanese prefectures are responsible for making industries comply with national environmental regulations. In the regions where national regulations are insufficient to achieve required standards, especially the highly industrialized areas like Tokyo Metropolitan Authority (TMA) area, the prefectures are given the authority to apply their own environmental standards through local ordinances. For example, TMA's strategy of controlling air pollution has three tiers: recommendation, local ordinance and contract. A recommendation is an administrative guidance asking industries to produce emissions only upto a certain percentage of emissions permitted by the national standards. The recommendation, which is normally followed by many industries for several years, generates information for the local authority about feasible abatement technologies and the cost of abatement for the industries, so that it can use that information in the design, promulgation and enforcement of local ordinance about environmental standards. The contractual agreement is a form of voluntary agreement, which local authorities make with companies involved in large-scale developments for making them comply with environmental standards tailored to specific developments. These agreements are related to cases in which either more stringent standards are required or there are no guidelines from the national standards.

The three-tiered regulatory structure is supported by the Environmental Council of Tokyo, an institution consisting of academics, officials of TMA and MITI and the Environmental Agency. TMA officials are well-informed on the technical issues relating to pollution abatement through technical hearings, in which industries testify under confidentiality on the current state and direction of technology. Public activism over environmental issues in the 1960s has resulted in the evolution of a locally based system for resolving pollution disputes. The MITI and the Ministry of Health and Welfare created their environmental pollution control divisions in 1963 and 1964 respectively. In 1965, the national parliament, Diet established a special committee for industrial pollution control. In 1967, Diet enacted the Basic Law for Environmental Pollution Control which set out broad policy objectives and priorities, outlined methods for implementing policy and defined responsibilities between central and local governments as well as between the public and private sectors. It also provided for the establishment of environmental quality standards (O'Connor, 1994). The National Law Concerning the Resolution of Environmental Pollution Disputes, 1970, provides guidelines for prefectures to establish environmental pollution investigative councils. These councils, by facilitating conciliation, mediation and arbitration, help prevent disputes from ending up in court.

In Japan, as in many other policy areas, the Diet plays only a minor role in the areas of environmental policy. Policy making is mainly the preserve of the bureaucracy. The bureaucracy is responsible both for formulation and implementation of environmental law. The Basic Law for Environmental Pollution Control, 1967, has established environmental standards for air, water, soil quality and noise levels. This led to the enactment of air, water and noise pollution control laws, the dispute settlement and health damage compensation laws and the Nature Conservation Law during the period 1968–73. However, the Basic Law, which forms the core of environmental policy, has set the tone for the relationship between government agencies, local authorities and industry, apart from providing a legal base for environmental policy.

MITI has mainly been responsible for implementing environmental legislation affecting industry. The guidelines issued by MITI on environmental performance under the laws on industrial promotion provide directions about the types of technologies that firms in a particular industrial sector have to use and develop in a particular timeframe. Firms normally follow MITI's guidelines, failing which, they face the threat of their non-cooperation being made public.

The Environmental Agency which was created in 1971 was given a horizontal role of coordinating environmental policies across the major ministries and creating environmental laws. It has developed a successful relationship with industry, facilitating a constant exchange of views and information between all relevant private and public institutions and representative bodies. In Japan, environmental rule making at the national level displays an interlocking set of processes in operation (Wallace, 1995). The main processes involved are the Environmental Agency's technical hearing system, which ensures that all relevant information is considered prior to legislation, a dialogue between MITI and industry, which allows MITI to determine optimal time scales for implementation and finally, a dialogue between the local authorities and industry which provides a direct route of feedback of public concerns to firms.

The New Basic Law for the Environment has placed sustainable development at the heart of environmental policy. It has created a legal base for a wide range of policy measures such as economic instruments. It has also provided for a common approach for environmental impact assessments and a basic plan for the environment at the national level. The New Basic Law has made the relationship between industry and government more transparent by providing a basis for voluntary agreements on pollution control.

MITI is mainly entrusted with the responsibility for the programme of voluntary measures to be undertaken by industry. When MITI called firms in 1992 to present proposals for voluntary pollution reductions, 60 per cent of major manufacturers submitted plans within two years for installing measures such as environmental audits, involvement of local communities, waste reduction and recycling. The voluntary agreement scheme at the national level builds on the already existing relationship between industry and the local authorities. At the national level, the voluntary agreements encourage firms to propose environmental measures, and consequently strengthen MITI's information base for setting guidelines or recommendations. As of 1991, there are some 37,000 pollution prevention agreements in effect in Japan (O'Connor, 1994). In a good number of cases, the agreements involved some form of local citizens' participation, with citizens reaching independent agreements with enterprises in some cases.

The attraction of voluntary agreements in Japan may be attributable to a number of factors (O'Connor, 1994):

- They provide a means for local governments and citizens to cope with new environmental problems associated with technology industries which, in some cases, are not adequately addressed by existing national regulations.
- Local governments are in a better position than the central government to devise control measures which are tailored to specific local conditions.
- Local residents and governments may desire to encourage the use of state-of-the-art pollution abatement technology whose development can outpace changes in national or local ordinances.
- Business may seek to improve their image among local residents through such agreements.

The Basic Plan for Environment, the requirement for which is laid down in the New Basic Law brings out the comprehensive and long run policies for environmental conservation and the matters required to promote these policies comprehensively and systematically. The environmental agency is responsible for drawing up the Basic Plan. The plan allows the local authorities in regions of serious pollution problems to draw up a local version of the Basic Plan. The local plan will be approved by the Prime Minister on the advice of the Conference on Environmental Pollution Control. Under the Basic Plan, local authorities retain their existing powers to set local environmental standards. However, limits are set by the Basic Plan on $CO_2$ emissions and other environmental concerns of national and global importance. The Basic Plan is supposed to create a national climate for voluntary measures by the industry. This nationwide recognition and adoption of voluntarism, now backed up by a coherent national strategy, build on the successful model of industry-regulator cooperation which has evolved over decades at the local level (Wallace, 1995).

Thus, Japanese environmental policy is mainly based on a participatory approach in which all the agents relevant for environmental management, government, industry and general public participate. The policy is built on the voluntary agreements between the government (local or national) and the industry to fix environmental standards and comply with them. Even though the New Basic Law provides for the use of economic instruments like carbon taxes, they are not used in Japan. However, as could be seen in the description above, the environmental policy of Japan is not guided by command-and-control (CAC) regulations either. The important government organization responsible for designing and implementing the environmental policy, MITI, plays only a catalytic role in making industry voluntarily agree to control environmental pollution.

## 1.5 LESSONS TO INDIA AND CONCLUSIONS

Indian government consists of central government, state governments and local self governments for cities, towns and villages. The revenue of state and central governments constituted around 24 per cent of GDP while the total government and public sector outlay formed 36 per cent of GDP during the year 1992–3.[3] The revenue receipts (including transfers from the centre) and outlays of state governments alone constituted 13.5 per cent and 18 per cent of GDP respectively during the same year. It has a big public sector supplying both private and non-marketable or public goods. The public sector constituted around 46 per cent of total capital stock in the economy and contributed almost one-third of gross of value-added in industry during the year 1992–3. The inefficiency of the public sector is reflected in a lower rate of return of 14 per cent on capital in comparison to a higher rate of return of 28 per cent in the private sector during the year 1992–3.[4] While India is comparable in terms of percentages of government revenue and government expenditure in GDP to advanced market economies including Japan, it differs

[3] See Economic Survey, Government of India, 1994 for details.
[4] See CSO, Government of India, National Accounts Statistics, 1995.

significantly from these countries in terms of expenditure allocations. A higher proportion of public outlays in India is market distortionary (outlays on the supply of private goods and general services) and a very low proportion of it is market-correcting (outlays on health, social security, environmental protection, etc.) as is normally found in many developing countries.

Public sector production of private goods, apart from contributing to efficiency losses through very low rates of return on investment in comparison to the private sector, has brought with it a variety of other distortions in the market through a number of supportive measures. These measures include reservation of products, preferential licensing, equity participation of government to keep the financial cost low to public enterprises, and the privilege of issuing tax-exempt bonds. Goldar (1996) provides a review of the role of government in India since Independence. The regulatory framework, consisting of a multitude of administrative rules and regulations and statutory legislations is another avoidable market distortion, introduced by the Indian government, with the objective of achieving certain social goals. These statutes and legislations include the Industrial Development and Regulation Act, Foreign Exchange Regulation Act, Monopolies and Restrictive Trade Practices Act, Import and Export (Control) Act, Capital Issues (Control) Act, Companies Act, Factory Act, Industrial Disputes Act, Trade Union Act, Minimum Wages Act, etc. The Indian government, being aware of the adverse effects of excessive controls and regulations in the economy, initiated the liberalization process at a slow pace since the mid-seventies and more actively by the mid-eighties. Subsequently, some of the statutes and legislations mentioned above have been repealed, and others have been made less stringent. The early nineties have seen major governmental reforms, covering all major areas of policy including trade policy, monetary policy and fiscal policy, aiming at a free market, with the gradual reduction of government control over production and distribution. If the current pace of reforms continues, the Indian economy may be a free market economy as some of the current West European or North American economies are, by the turn of the twentieth century.

Ironically, while intruding in a big way since Independence, into areas where the market can function effectively for achieving the stated goals of economic development, the Indian government has not bothered to deal with the acknowledged areas of market failure such as the management of environmental resources until the mid-seventies. Most of the environmental legislations came into existence only during the seventies and eighties. These legislations have contributed to the creation of necessary institutions at different levels (centre and state) of government to prevent and monitor environmental pollution. In spite of having these institutions, India is lagging far behind in the design and enforcement of economic instruments such as pollution taxes and marketable pollution permits. Recent research on water pollution abatement in India has shown that the rate of pollution tax required to realize the prescribed standards must be much higher than the currently levied water cess. However, some recent studies on industrial pollution abatement in India (Mehta, Mundle and Sankar, 1993; James and Murty, 1996) have dealt with the problem of using pollution taxes for controlling water pollution by Indian industry. These, including other ongoing studies, may provide necessary inputs to the government for using economic instruments in pollution control. Murty (1995a) provides a review of the current status of environmental regulation in India. There is now a growing realization in India that there are areas where economic instruments or CAC type of regulations do not work and there is a need for collective action *a la* Coase and Becker, as described in Misra and Murty (1995) and Misra (1996).

The role of the government in India was not complementary to the market until the mid-eighties. There has been a neglect of development of social sectors which supply merit wants type of services, management of environmental resources which are public goods and supply of public utility services which a free market alone cannot efficiently handle. The economic reforms in early nineties for liberalizing the economy have rightly started the process of reducing governmental control in areas where the market is efficient, but has not augmented the needed government intervention in the areas of market failure. For making necessary changes in its economic policy, India has to draw from the experiences of the more mature market economies of Western Europe and Japan.

The Indian government faces the problem of choice of instruments for achieving the developmental or non-developmental objectives. It can achieve many of these objectives either by direct intervention in the market process or by playing a catalytic role. The experience of Japan discussed earlier in Section 1.4 provides an example of the catalytic role of the government in dealing with both economic development and market failures. India has to learn a lot from the Japanese experience in this area.

India currently has a top-down approach in its environmental policy, which has been ineffective so far. In contrast, the Japanese environmental policy is based on a bottom-up approach, in which the local governments, industries and communities have played important roles in shaping the nation's environmental policy. India, being a federal country with many levels of government, should give more autonomy to local governments to deal with environmental problems. Some recent studies (Chopra, Kadekodi and Murty, 1990; Misra and Murty, 1995a; and Misra, 1996) have shown that there are many cases in which local governments, NGOs, local communities and industry can collectively solve the environmental problems. A successful environmental policy in India requires the support of local initiatives, with government (central or local) playing a catalytic role.

# APPENDIX A 1

Table A 1.1
Central Government Expenditure by Type in Industrial and Developing Economies, 1986–7
(Per Cent of Total Expenditure in 1986–7)

| Region | General Public Services | Defence | Education | Health | Social Security | Transport & Commu-nication | Other Economic Services | Others | Central Expenditure (% GDP) | Central Revenue (% GDP) |
|---|---|---|---|---|---|---|---|---|---|---|
| Industrial | 8.05 | 7.34 | 8.45 | 9.61 | 37.71 | 5.12 | 7.15 | 16.57 | 31.46 | 26.99 |
| Developing | 16.94 | 11.97 | 14.11 | 6.21 | 12.60 | 7.19 | 16.04 | 14.94 | 25.40 | 21.10 |
| Africa | 18.54 | 8.85 | 15.97 | 5.51 | 8.25 | 7.12 | 17.49 | 18.27 | 25.63 | 21.23 |
| Asia | 17.87 | 12.99 | 14.39 | 5.18 | 7.14 | 11.23 | 19.41 | 11.79 | 19.89 | 17.25 |
| Europe | 16.88 | 13.63 | 7.26 | 6.26 | 23.07 | 6.53 | 21.18 | 5.19 | 29.15 | 28.46 |
| Middle East | 13.29 | 26.69 | 12.54 | 4.85 | 13.22 | 3.71 | 11.06 | 14.64 | 33.12 | 25.82 |
| Western Hemisphere | 16.09 | 7.53 | 13.98 | 8.32 | 18.77 | 6.80 | 11.88 | 16.63 | 25.04 | 19.64 |

*Source:* Burgess and Stern (1992).

Table A1.2
Water-Polluting Industries

1. Ferrous metallurgical industry
2. Non-ferrous metallurgical industry
3. Mining industry
4. Ore-processing industry
5. Petroleum industry
6. Petro-chemical industry
7. Chemical industry
8. Ceramic industry
9. Cement industry
10. Textile industry
11. Paper industry
12. Fertilizer industry
13. Coal (including coke) industry
14. Power (thermal and diesel) generating industry
15. Processing of animal or vegetable products industry

*Source:* The Water (Prevention and Control of Pollution) Act, Government of India, 1974.

Table A 1.3
Air Polluting Industries

1. Asbestos and asbestos products industries
2. Cement and cement products industries
3. Ceramic and ceramic products industries
4. Chemical and allied industries
5. Coal and lignite based chemical industries
6. Engineering industries
7. Ferrous metallurgical industries
8. Fertilizer industries
9. Foundries
10. Food and agricultural products industries
11. Mining industry
12. Non-ferrous metallurgical industries
13. Ores/mineral processing industries including beneficiation, palletization, etc.
14. Power (coal, petroleum and their products) generating plants and boiler plants
15. Paper and pulp (including paper products) industries
16. Textile processing industry (made wholly or in part, of cotton)
17. Petroleum refineries

(*Cont.*)

18. Petroleum products and petro-chemical industries
19. Plants for recovery from and disposal of wastes
20. Incinerators

*Source:* The Air (Prevention and Control of Pollution) Act, Government of India, 1981.

Table A 1.4
Rates of Water Cess

| Sl. No. | Purpose for which water is consumed | (Paise per kilolitre) Rate | Rate under sub-section 2A* |
|---|---|---|---|
| 1. | Industrial cooling, spraying in mine pits or boiler feeds | 1.50 | 2.25 |
| 2. | Domestic purpose | 2.00 | 3.00 |
| 3. | Processing whereby water gets polluted and pollutants are easily bio-degradable and are toxic | 4.00 | 7.50 |
| 4. | Processing whereby water gets polluted and pollutants are easily bio-degradable and are not toxic | 5.00 | 7.00 |

*Source:* The Water (Prevention and Control of Pollution) Cess Act, 1977.

*Note:* * Under Section 25, sub-section 2A, of the Water (Prevention and Control of Pollution) Act 1974, a person cannot establish an industry which is likely to discharge sewage or effluent into a stream or well without the prior consent of State Boards. If a person can establish industry under certain provisions then the rate of cess shown in column 3 will apply.

# 2. Using Environmental Policy Instruments: A Review of International Practice

A. J. JAMES

Although governments of different countries have sought to deal with environmental problems in several ways, the performance of policy instruments varies widely depending on the context of their use and the magnitude of the problem. This chapter provides an overview of these policy instruments and reviews their performance across various countries. The main lessons from this exercise for developing countries in general, and India in particular, are (a) that policy instruments cannot be transplanted irrespective of the context of their intended use, and must be selected and modified with care and judgement; and (b) that the actual use of environmental policy instruments must be phased-in, so as to maximize the chances of both, effective pollution control and political acceptability.

This chapter is divided into three sections. Section 2.1 evaluates the performance of these instruments across various countries, the following section discusses some issues relating to the implementation of environmental policies, and Section 2.3 concludes.

## 2.1 ENVIRONMENTAL POLICY INSTRUMENTS FOR INDUSTRIAL WATER POLLUTION CONTROL: THE INTERNATIONAL EXPERIENCE

### *Northern America: The USA*

*Legislation:* Environmental regulation of the discharge of waste into American waters started with command-and-control (CAC) measures to control point-source pollution. The Water Pollution Act of 1956 provided for (1) federal financial support for constructing wastewater treatment plants and (2) direct federal regulation of waste discharges (Tietenberg, 1986). The 1972 Amendment considerably toughened the law on discharge of wastes into water bodies and enforced a regulatory regime; permits issued by the Environmental Protection Agency (EPA), which were granted only when the dischargers met certain (uniformly imposed) technology-based effluent standards were required of all dischargers. There were two general policy goals: (1) an ultimate objective—to eliminate all discharges of pollutants into navigable waters by 1985; and (2) an interim 'fishable-swimmable' goal—to protect aquatic life and recreational uses of water bodies (Oates, 1985). The effluent standards themselves were to be implemented in two stages: by 1977, all industrial dischargers had to meet effluent limitations based on the 'best practicable control technology', and by 1983, they had to meet more stringent 'best available technology economically achievable' standards (Tietenberg, 1986). Also, the federal government set for itself the major role of funding the construction of municipal waste-treatment plants, paying upto 75 per cent of the cost of these plants, which ran into several billions of US dollars per year (Oates, 1985).

Following the 1974 Safe Drinking Water Act, which set stringent standards for community water systems (in terms of maximum allowable concentration levels for bacteria, turbidity and chemical-radiological contaminants), the 1977 Amendments to the Water Act continued with this regulatory approach, but with some major modifications. A more careful distinction between conventional and toxic pollutants was made, with more stringent requirements placed on the latter, and it extended all the deadlines of the 1972 Act (Tietenberg, 1986). The 1977 Amendment also introduced compulsory pre-treatment, upto certain standards, for waste being sent to a publicly owned treatment system. These standards were set to safeguard the municipal treatment plants and to ensure that toxic pollutants would not be introduced into the waste-treatment facility.

*Effluent Fees:* There is a scattered use of effluent fees for controlling water quality in current US environment policy. There was an initial positive feedback on the effectiveness of fees for municipal and industrial wastewater on reducing pollution, and fostering investment and innovation in water pollution control technology (Oates and Baumol, 1975). Oates and Baumol (1975) give examples from Kneese and Bower (1968), and Fisher (1956) to illustrate early successes in reducing industrial emissions through a charge. Kneese and Bower cite three instances in which levying local sewer charges induced striking reductions in waste discharges, while Fisher describes the case of a fee established in 1953, with the proviso that anyone who could meet a specified set of standards by a certain date would get a rebate; 23 major companies invested five million dollars on pollution control in less than two years in order to meet these standards. Oates and Baumol (1975) also mention the work of Elliot and Seagraves (1972), who show that there is econometric evidence to show that industrial BOD emissions and water consumption appear to respond negatively to the level of surcharges on emissions. They accordingly conclude on an optimistic note: 'We are thus beginning to accumulate some evidence indicating the effluent fees can, in fact, be quite effective in reducing levels of industrial waste discharges into waterways' (ibid, p. 118).

But fees levied on industrial emissions into municipal waste-treatment plants, although based on both the quantity and the quality of the effluent, are not based on actual emissions. They are instead based on 'average' levels of discharge, and are primarily revenue-generating. The current view therefore, is that '[t]heir role as an economic incentive to regulate levels of emissions has apparently been minor' (Cropper and Oates, 1992, p. 395, fn. 20).

*Marketable Permits:* The most celebrated of the marketable permit schemes covering water pollution in the USA is the Tradable Discharge Permit (TDP) scheme on the Fox river in Wisconsin (Tietenberg, 1986). Although the centre piece of American regulatory reform in water pollution control, it has not been as high-profile as emissions trading[1] for air quality control. The permit scheme was to cover 10 paper and pulp mills and 4 municipal waste-treatment plants discharging waste into a 22-mile segment of the lower Fox river flowing from lake Winnipeg to Green Bay. The problem leading to the establishment of the permit scheme was that, due to the pollution, dissolved oxygen (DO) standards were not being reached at two critical points, even though discharges were in compliance with the standards. A study showed that control costs would be 40 per cent higher if traditional abatement rules were used and emissions trading possibilities were denied, and that there were potential annual savings from a permit scheme estimated at

[1] Emissions trading refers to sale and purchase of marketable permits.

6.7 million US dollars (*ibid.*). Allowable emissions were also varied with the river's seasonal capacity to assimilate wastes, and short-term leasing of permits was allowed (Oates, 1985).

In March 1981, transfer of permits was allowed and the first trade was carried out: 'a paper mill which shifted its treatment activities to a municipal wastewater treatment facility' (Cropper and Oates, 1992, p. 690). But that remains the sole trade. Cropper and Oates feel that trades were apparently discouraged by a set of quite severe restrictions, and that the conditions needed for the emergence of a viable market in discharge permits had not been established. However, emissions trading has not been taken up by any other state in the USA and the only other such notable scheme is that of phosphorous discharge into the Dillon reservoir. But no trades had taken place by 1992 (*ibid.*). Thus Cropper and Oates (1992) conclude that the US experience with tradeable permits is 'a limited one with quite mixed results' (p. 691).

*Water Quality in the USA:* Some striking successes in water quality management have been noted in the USA, including the return of Atlantic Salmon to Connecticut and Penobscot rivers in New England and the revival of lakes Erie and Ontario. But the general feeling is that there has been little real improvement in surface-water quality over the decade 1975–85; and that there has been no significant change in the concentration of DO, suspended solids (SS), dissolved solids (DS), faecal coliform bacteria and phosphorous in surface-water between 1974 and 1981 (Oates, 1985).

However, considerable progress has been made in the control of point-source discharges: industrial discharges fell by 70 per cent or more between 1972 and 1977; and the quantity of oxygen-demanding pollutants emitted by municipal treatment plants fell by 46 per cent, despite a 12 per cent increase in pollutants entering these plants (*ibid.*). The problem areas of non-point source pollution and contamination of groundwater however remain to be solved.

### *Europe*

All the environmentally-progressive OECD countries started off with a regulatory framework; economic instruments have only been used at a later stage. The experience of OECD countries suggests that effluent charges have been far more popular as economic instruments than marketable permits, in contrast to the experience of the USA. But some OECD countries (like Germany and the

Netherlands) are observing the performance of US emissions trading programmes as a possible replacement for the existing CAC systems. The water pollution control measures and experiences of some selected OECD countries are discussed below.

*Germany:* Environmental policy in some parts of Germany dates back to the nineteenth century, when (and until the Second World War) environmental policy was known as 'air and water hygiene' (Wallace, 1995, p. 63).

The first river authority (*Genossenschaft*) in Germany was organized along the Emschen river in the Ruhr valley in 1901 to treat industrial waste. It was financed by effluent charges on their members (Oates and Baumol, 1975). By 1975 there were 8 *Genossenschaften*, which together form a 'closed water-control system which has maintained a remarkably high quality of water' in the eight rivers they control. In fact, in all but one of the rivers, water was suitable for aquatic life and swimming.

The fees collected (in 1975 US dollars) amounted to about 60 million a year, mainly from effluent charges levied on their nearly 500 public and private members (in 1975), with the levels of charges based largely on a set of standards for maintaining water quality. The point to note is that the charges themselves were based on rather complicated formulae which, however, do not represent marginal-cost pricing (*ibid.*). German effluent fees, like many others in Europe, are not primarily intended to regulate industrial discharges, but rather to raise funds to finance projects for water quality management. Fees are therefore typically low and applied on 'average' or 'expected' discharges, and not on the actual discharge (Cropper and Oates, 1992). Also, charges are overlaid on an extensive CAC system of regulations. Dischargers are required to meet minimum standards for wastewater treatment for a number of defined pollutants. Simultaneously, a fee is levied on every unit of discharge depending on the quantity and noxiousness of the effluent. However, there is a two-part tariff, with dischargers meeting or exceeding the effluent standards liable to pay only half the normal rate (Tietenberg, 1991). Apart from rate of effluent charges being low, the underlying regulatory system further dilutes any effect they may have had on the economic incentives facing a water-polluting firm to innovate or invest in pollution control technology or even to reduce output. Germany also provides subsidies to facilitate the implementation of environmental programmes and to help small firms with cash flow problems to install pollution control equipment (EPAT, 1993).

*France:* Like Germany. France has a system of regulations for the control of water pollution, overlaid with a network of emission charges for the discharge of industrial effluents into water bodies. Again like Germany, these charges are not high enough to act as economic incentives for firms to innovate or invest in more effective pollution control technology, and are based on 'average' or 'expected' levels of discharge and not on actual levels. They are thus primarily aimed at raising revenue for the government to invest in public programmes to improve inland water quality (Cropper and Oates, 1992).

The French legislated water pollution charges in 1970, aiming to charge polluters with the full social cost of pollution. Municipalities were expected to run wastewater treatment systems at the local level, while a higher level of six 'financial river basin agencies' (AFBs), whose areas correspond largely to the catchment areas of the six biggest rivers in the country, are responsible for the management and development of the entire service (Scott and Lawlor, 1994).

The municipalities contract with industries to treat the latter's effluent and obtain grants for pollution abatement from both the AFBs and from the central government. The AFBs levy a pollution charge on all dischargers, and use the charges primarily to fund the wastewater system and only incidentally as an incentive to reduce pollution. The revenues are thus disbursed as pollution abatement grants and loans to both, municipalities that operate the public water system and industries discharging directly into the inland waters.

The AFB's charges are based on COD, suspended solids, nitrogen, phosphates, toxic materials and dissolved salts in the effluent, and are calculated in terms of francs per kilogram per day. In areas where the pollution problem is acute, the charge is increased by multiplying it by a coefficient (Jansen, 1991). Charges, however, are low and this is reflected in the low levels of municipal treatment and in the fact that the system is not self-financing. Still, industry has reduced its pollution levels by 70 per cent since the introduction of the effluent charge system, though the non-complying industries are generally the ones that generate the more toxic pollutants (*ibid.*).

Apart from the effluent charges and the resultant subsidies, the French government also supports technological development as a means to achieve environmental goals. In 1990, an organization called the ADEME (*L'Agence de L'Environnement et de la Maitrise de L'Energie*) was created to fund and co-ordinate research and development on environment technologies. Such research seeks to develop new technologies not only for domestic industry, but also for export to other countries.[2]

*The Netherlands:* The Netherlands has one of the oldest and most effectively managed systems of charges and also one with relatively high fees (Cropper and Oates, 1992). There is some evidence to suggest that this fees has caused firms to respond with significant cutbacks in discharge of water-pollutants and have thus had a significant effect in reducing water pollution. Once again, however, as in Germany and France, these measures apply to point-source pollution and not to non-point source pollution or groundwater pollution (*ibid.*).

The Netherlands passed its Act on Pollution of Surface Water in 1970, which divided responsibility for water quality management between the state and regional authorities (Scott and Lawlor, 1994). The bulk of its water resources—major rivers, lakes, and coastal waters—were deemed to be 'state water surfaces' and became the responsibility of the Ministry of Public Works and Transportation and Environmental Protection. 'Regional water surfaces', corresponding roughly to watershed areas, were the responsibilities of regional authorities, which were responsible for collective water treatment, water supply undertakings and sewerage networks. Regional authorities issue licences for discharges to surface waters and sewerage, specifying any pre-treatment deemed necessary, and the quantities of pollutants to be discharged (*ibid.*).

[2] This is a strong illustration of the 'revisionist' view of environmental regulations popularized by Michael Porter of the Harvard Business School. This view asserts that 'tough regulations will stimulate innovation, making firms generally fitter and more competitive', and states that strict environmental standards can lead to national competitive advantage, either by (1) forcing its industries to develop improved processes or dedicated pollution-control equipment, which it can then sell to other countries who subsequently adopt similar standards; and (2) by stimulating industrial innovation in devising superior technologies and improving corporate performance that bring competitiveness benefits 'which outweigh any additional costs of meeting the increased standards' (Wallace, 1995, p. 33).

The system is self-financing, at the regional and the national levels. The main stated aim is to finance water purification. A subsidiary aim is to give polluters an incentive to lessen pollution. However, the pollution charge is not based on the marginal-cost pricing principle, and instead is based on levels of COD and nitrogen, and (now) also on quantities of heavy metals. Medium-sized enterprises are charged on the basis of both their levels of production and their average pollution levels. Discharge to state waters is also priced and regional authorities have to pay for such discharges (*ibid.*).

Since 1970, there has been a significant reduction in industrial pollution, which was larger than anticipated, and resulted in an overcapacity in the system. This in turn led to a 500–800 per cent upward revision in annual charges (in nominal terms) over the period 1970–90. Most of the funds collected through these charges are spent on capital investment, with monitoring of discharges accounting for only a small proportion of costs, since only random checks are carried out. Grants are also available to firms for installing pollution abatement plants for water treatment before discharging into state waters. But it is envisaged that these grants will be phased out by the end of the century (Scott and Lawlor, 1994).

An area of environmental policy which has had a significant effect on pollution levels in the Netherlands is that of voluntary agreements with industry, where industry volunteers (readily or reluctantly), to reduce emissions and is made to stick to that agreement. Results have been surprisingly good (Wallace, 1995).

*Denmark:* Although Denmark has a record of high environmental standards, until recently it had no effluent charges because they contrasted with the prevailing Danish consensus-seeking policy style (Anderson, 1994). Instead, the State undertook to clean all industrial effluents by constructing public treatment plants and from 1974 to 1987, water pollution control was rarely at source, but basically by end-of-pipe solutions.[3] By not charging industry the true costs of waste water treatment, the Danish water pollution control policy has been 'both expensive and ineffective' (*ibid.*).

Only the city of Copenhagen provides an exception to this rule. From 1979, effluent charges were imposed and there was a

[3] End-of-pipe solutions refers to technologies which clean dirty industrial waste (water or air) *after* the industrial production process has been completed. The contract is with 'clean(er) technologies', which ensure that the *production process itself* does not generate excess pollutants.

considerable reduction of effluent-load as a result of firms adopting cleaner processes in production technology. Specifically, the Copenhagen authority imposed a sewerage charge on companies and a surcharge to cover treatment of the trade effluents which were below the average quality of sewage. A 20 year payback period was chosen (1979–99), after which the capital part of the surcharge would end, and industrial customers would have to pay only the operating surcharge. But payment being on a volumetric basis (and not upfront on the basis of capacity reserved), when firms adjusted their pollution emission loads downwards, the income to the authorities fell correspondingly. The deficit that arose had then to be paid by all customers, including households. Those who had found themselves with higher payment burdens, however, could now enjoy the twin benefits of a self-financing wastewater treatment plant and significantly reduced levels of industrial pollution (Scott and Lawlor, 1994).

Firms that started the scheme early, also have a higher payment burden than newer entrants, who have to pay the surcharge for a fewer number of years. But their consolation is that the system, while seemingly unfair, is administratively simple. This is symptomatic of a system characterized by a high degree of interaction between government and industry, or more specifically, between Danski Industri—the main representative body for Danish industry—and the Pollution Ministry (Wallace, 1995). In fact, the Danski Industri was closely involved in the formulation of the Pollution Act of 1973.

Since 1992, the primary legislation in Denmark is the Environment Protection Act, and Danish environmental policy is the responsibility of the Environment Ministry and is administered through the Environmental Protection Agency (the successor to the original Pollution Ministry) (*ibid.*). The latter has five main areas of activity, which are listed below:

1. Approving industrial operations, including identifying polluting or hazardous activities, producing guidelines for dealing with these activities, and assigning responsibility for applying the guidelines, all of which are negotiated with industrial organizations and concerned organizations in advance.
2. Supporting cleaner technologies in an attempt to regulate the emission of heavy metals (when it was realized that end-of-pipe treatment would be extremely costly).

3. Formulating agreements with industry on the same lines as the Netherlands.

4. Providing economic incentives with an extensive though newly-developed system of pollution charges and fees, especially for waste-treatment and disposal, used primarily for revenue-generation with revenues being earmarked for development and demonstration of cleaner technologies.

5. Providing infrastructure, by centralizing waste-treatment facilities for industries, funded locally by municipal authorities, which firms are obliged to use for their waste-treatment and waste-disposal needs (*ibid.*).

*Italy:* The Italian charge system for industrial water polluters is mainly designed to encourage polluters to achieve the provisional effluent-treatment standards as soon as possible. To this end, the effluent charge is nine times higher for firms that do not meet the prescribed standards than for firms that do. Also, the charge system being designed only to facilitate the transition to the prescribed standards from the provisional standards, it was scheduled to expire, once full compliance was achieved. This was initially set at 1986, but has been postponed since (Wallace, 1995).

### *Asia*

The Japanese case, which has been dealt with in detail in Ch. 1, represents one of the most striking cases of rapid and effective control of air and water pollution problems in Asia, with a strong regulatory framework being effectively complemented by suasive instruments like voluntary pollution control agreements, and economic instruments like subsidies on pollution control technologies. A focus on research and development in pollution control and cleaner technologies has also stood them in good stead on the international market. The main reason for the effectiveness of these measures is the close historical link between industry and government, especially in policy-making.

The Japanese case provides a role model for other Asian countries, and Korea, Taiwan, Indonesia and Thailand have followed suit with environmental policies to tackle problems of industrial pollution. But there are variations in the choice of policy instruments and the relative emphasis across the range of instruments varies, reflecting the local needs and context of environmental policy in each of these countries. A brief account of the environmental policy framework and effectiveness of the enacted measures in each of these five countries follows.

*Japan:* Japan had an early taste of environmental pollution during the early industrialization drive from 1868 to 1945, when mining operations caused widespread damage to farmland (Wallace, 1995). Growing public protests in the 1950s galvanized local government initiatives to curtail industrial pollution problems, and the national government responded with the Basic Law for Environmental Pollution Control in 1967.[4]

In the early seventies Japan enacted some of the strictest environmental standards in the world, backed with strong regulatory enforcement, in the face of an escalating problem of industrial waste-water and air pollution during the second industrialization drive of the fifties and sixties (O'Connor, 1993). These CAC measures effected a rapid and substantial decrease in the ambient levels of many pollutants by the late seventies and early eighties. Such regulation continues to form the cornerstone of Japanese environmental policy and poses a strong counter-example to the theoretical view that market-based instruments are more effective than regulatory measures. The main lesson from one of the oldest and most highly developed CAC systems is that 'once government priorities have incorporated environmental concern, much can be accomplished in a relatively short time-frame through an effective command-and-control (CAC) system' (*ibid.*, p. 34).[5] The more remarkable facet of Japan's environmental policy is that Japanese economic growth continued at around 6.5 per cent per annum during the seventies and 4 per cent per annum during the eighties (*ibid.*) Given the close co-ordination between government and industry, Japan's regulatory approach is effective, dynamic and oriented towards cleaner technology rather than to conventional end-of-pipe treatment.

[4] See Wallace (1995, Ch. 6) for a detailed account of early environmental pollution problems in Japan and the consequent public and policy response.

[5] Japan's environmental laws provide for the use of command-and-control measures. However, as explained in Ch. 1, in reality environmental pollution is efficiently handled by the co-operation of local governments, communities and industry so that the need for using the legislated CAC measures rarely arises in Japan.

Another policy instrument used extensively by the Japanese government is an environment fund, primarily financed by fines on polluters, and used initially to compensate victims of environmental pollution. Created by the Pollution-related Health Damage Compensation Law in 1974, and intended to compensate victims of specific pollution-related illnesses like the Minimata, Itai-itai and air pollution-related asthma, the general improvement in environmental quality has recently led the government to stop accepting new compensation claims.

Also used widely in the Japanese case are pollution control agreements between industry and government or local communities. These are usually not legally binding; they rely on moral suasion and self-policing, and are intended to alter polluting behaviour within industry. Normally, such agreements specify what measures a firm is willing to take in order to prevent or minimize pollution, and also specify a time period for their introduction. The strongest incentive for companies to effect such voluntary compliance is the impact such a policy has on public relations with the local community. The fact that, by 1991, there were some 37,000 agreements in force, with 259 of them being reached without local government involvement (O'Connor, 1994) indicates the effectiveness of such a policy instrument.

A strong policy to foster innovations in pollution control policy has not only helped domestic industry to abate pollution quickly and efficiently, but has also made Japan a leading world supplier of cleaner technologies and pollution-control equipment. The initiative began with the establishment of the Environmental Pollution Control Service Corporation, since re-named the Japan Environment Corporation (JEC), which also provided low-interest loans for small-scale firms to install pollution control facilities (*ibid.*). Subsidies are also accorded in the form of tax concessions for pollution control and waste-treatment facilities constructed with the help of the JEC. Japan is also notable for not having distortionary pricing policies, that is, policies that foster environmental pollution or provide economic disincentives to invest or innovate in pollution abatement technologies. However, there are no tradeable permit schemes in force in Japan (*ibid.*).

A new law, the Basic Environmental Law of 1993, reflects the changes in industry-government relations during the late 1980s and early 1990s, and gives adequate emphasis on the imperative of sustainable development. It not only creates a legal base for a wide range of policy measures such as, education and financial instruments, but also underscores the need for a nationwide implementation of environmental impact assessments, reflecting the new long-term perspective on environmental performance (Wallace, 1995).

*Korea:* There are several similarities in the nature and performance of environmental policies in Korea and Japan. Like Japan, Korea also relies primarily on CAC policies. One reason for its effectiveness in controlling industrial wastewater pollution is the nature of its industrial structure, with a small number of very large firms (the *chaebol*) accounting for a major share of industrial output and industrial pollution (O'Connor, 1993). The possibility of targeting the worst offenders therefore represents relatively low monitoring costs. Like Japan, the close government-industry consultations that characterize environmental policy making in Korea, facilitate the adoption of a dynamic process-oriented technology reform, instead of the usual focus on end-of-pipe treatment (*ibid.*).

Korea also has an Environment Pollution Prevention Fund, financed in part by the government, and in part by fines or pollution charges levied on firms exceeding the emission standards laid down in the regulatory framework. The penalty rates in this taxes-and-standards system, however, are not high enough to provide an economic disincentive to pollute above the standards. Attempts to raise the rates have not been successful, as firms have been pleading financial hardship. Consequently, the accent is on providing low-interest loans for the installation of pollution control technologies, besides compensating victims of environmental pollution. But while there are no tradeable permit schemes in operation in Korea, like Japan, one difference is the limited role of voluntary pollution control agreements between industry and government or local communities.

*Taiwan:* General environmental legislation in Taiwan has its origins in the Public Safety Law of 1971 and its culmination in the Environmental Protection Fundamental Law of 1988, after a revision in 1982 (Chiang, 1994; Shaw, 1994).

Regarding water pollution and control, Taiwan announced a Water Pollution Control Act in July 1974, which it revised twice since then, the latest being in May 1991. The latest Act calls for (i) 'technology-based and economically feasible' standards for ambient environmental quality; (ii) the construction and management of public and private sewer systems and wastewater treatment facilities, co-ordinated with water pollution control laws and policies; (iii) regular monitoring of industrial water quality; and (iv) an increase in the fines for defaulting industries, by at least ten times the previous limits, coupled with prosecution of industries who refuse to comply with government orders to shut down (Chiang, 1994).

In order to prevent the proliferation of new pollution sources, the Ministry of Economic Affairs promulgated the 'Guidelines for Evaluation of Pollution Control in New Plants' in 1988, which classifies industries into two types: Class A, which have to conduct an environmental impact assessment exercise before they are licensed to start production, and Class B industries which have to submit proposals for pollution control for government scrutiny before licences are issued (*ibid.*).

Although Taiwan has a regulatory framework for the control of environmental pollution and setting down standards for ambient environmental quality, it is more inclined to experiment with the use of economic instruments. The main reason is that, unlike Korea and Japan, the market structure is not so highly concentrated and there are a larger number of geographically-dispersed small-and medium-sized enterprises. This raises the costs of monitoring and would not make a straight CAC approach practicable. Taiwan is currently studying the feasibility of introducing pollution taxes to control water pollution in the country (O'Connor, 1993). In addition to regulations governing the discharge of water-borne pollutants, Taiwan also provides technical assistance for industrial waste-minimization and for industrial pollution control. Experts, scholars and engineers contribute to the Technical Services for Industrial Pollution Control, which provide technical assistance for pollution control to industries (Chiang, 1994).

Economic instruments currently in operation in Taiwan consist mainly of (i) import duty exemptions for imported pollution control equipment; (ii) tax reductions on investment in automated facilities for technology, pollution control technology, research and development and training costs; (iii) accelerated depreciation on pollution control equipment; (iv) income-tax exemption for shareholders of newly-issued ordinary shares, for the same fiscal year; (v) encouragement to relocate industries to better-planned industrial zones; and

(vi) low-interest loans for the installation of pollution control equipment within the private sector, and preferential loans (from the Sino-American Fund) for strategic industries, small and medium businesses, and promotional loans for the guidance of small and medium businesses (*ibid.*).

According to the statistics from the Environmental Protection Agency of Taiwan, 28 per cent of the investments in pollution control were for controlling industrial wastewater pollution. 32 per cent of applications for manufacturing industries for exemption from import duties came from wastewater treatment projects, while the figure was 17 per cent when applications from all industries are considered. Among applications for investment tax deductions, nearly 44 per cent came from wastewater treatment units (*ibid.*).

Here again there are no tradeable permit schemes in the country and voluntary pollution control agreements are also quite limited (*ibid.*). There are however plans to adopt an economic incentives approach to pollution treatment, including an emission fee system, a total emission control system for water and air pollution, and an offset system for air pollution (Shaw, 1994).

*Indonesia:* Indonesia has perhaps the newest system for environmental pollution control, and is currently in the process of establishing a basic regulatory framework. In 1989, a comprehensive environment bill was made into a law, one section of which 'establishes a national policy for maintaining reasonable standards of purity for the air and water of the country' (Prabowo, 1994). Pollution control in Indonesia is primarily the responsibility of the National Pollution Control Commission, an inter-departmental body consisting of agencies from various Ministries (*ibid.*).

The government prefers to concentrate initially on building monitoring and enforcement capabilities before considering the use of economic instruments (O'Connor, 1993). Nevertheless, given that Indonesia has modelled its environment management system after Canada, which is essentially of the CAC type, it may well be that Indonesia will assign a less important role to economic instruments like effluent charges. The other consideration of course is that being a low-income country the government is averse to measures which would impose additional cost burdens on consumers (*ibid.*).

The only formal regulatory programme with respect to water pollution is exclusively addressed to industrial wastewater pollution, which is implicit in the factory licensing function of the Ministry of

Industry. The Ministry can order modifications or suspend or revoke these licences if it feels that sufficient care has not been taken in the disposal of water-borne wastes. Nevertheless, despite the provision of substantial fines and even prison sentences in case of violation of the orders of the Ministry, enforcement is sporadic (*ibid.*).

Subsidies continue to be a popular economic policy option. The Ministry of Population and Environment gives concessions on import tax duties for wastewater treatment equipment, and donor agencies are also considering the possibility of giving subsidies for the installation of wastewater treatment facilities for certain industrial sectors, and for the adoption of cleaner technologies (*ibid.*). The government is also considering the provision of common wastewater treatment facilities for small-and medium-scale industries.

Although marketable permit schemes are non-existent, there is widespread use of pollution control agreements between industry and the government. As explained earlier, in the case of Japan, such agreements are not legally binding and rely on moral suasion and self-policing to change polluting behaviour. In Indonesia, there are two well-focused programmes called PROKASIH and PROPER[6], which are aimed at cleaning up the most polluted rivers. Under the PROKASIH programme initiated in 1989, major pollution sources along the 20 most polluted rivers were targeted, and firms with the heaviest pollution loads were asked to draw up pollution abatement plans on which the agreements were to be based. Initially, between 1990 and 1994, there was a drop in BOD levels in the rivers, but in recent years they began rising again (O'Connor, 1996).

In order to counter this reverse, the government initiated another programme called PROPER, which sought to counter the negative publicity focus of the PROKASIH programme. Instead of relying on public pressure generated by the publication of the worst offenders, the government now regularly publishes information on environmental ratings, drawn up on the basis of pollution performance. Industries which meet international standards of effluent discharge are termed 'Gold' industries, while those which meet national standards are termed 'Green'. 'Blue', 'Red' and 'Black' industries represent worsening levels of industrial performance on pollution control. Although no firm has met the 'Gold' standard, between June 1995

[6] PROKASIH and PROPER': Translated from Indonesian. They basically mean Clean River Programmes.

and December 1995, there was a discernible movement from 'Black' to 'Blue' standards: the number of 'Black' industries went down from 6 to 3, the number of 'Red' industries went down from 115 to 108, while the number of 'Blue' industries went up from 61 to 72 (*ibid.*). It would thus appear that 'positive' publicity coupled with 'negative' publicity on industrial pollution performance is a better configuration than 'negative' publicity alone.

*Thailand:* Although Thailand has a regulatory framework like most Asian countries, it is inclined like Taiwan towards experimenting with the use of economic instruments (O'Connor, 1993). It is actively considering the imposition of pollution taxes. The reason for this shift away from CAC measures is the nature of its industrial structure, which tends to be more geographically dispersed as in Taiwan, and has more small-and medium-scale industries than either Korea or Japan.

Among the economic incentives currently being offered for pollution-abatement to polluting industries, is a discount by the Ministry of Finance on the standard tariff on imported capital equipment for end-of-pipe waste-treatment technology (*ibid.*). However, this does not have a large effect on the import and installation of waste treatment technology, with the value of imports on which duties had been reduced being only around 7 million US dollars by 1989. In fact the government is considering extending this discount to the import of cleaner process technologies as well (*ibid.*). In addition, the semi-governmental organization, the Industrial Finance Corporation of Thailand (IFTC), offers concessional financing of anti-pollution equipment purchases, and lends around 8 million US dollars annually through a recently-established Environment Portfolio (*ibid.*).

Thailand also launched an Environment Fund, on the same lines as those of Japan and Korea, with an initial government capital contribution of around 20 million US dollars, which is to be supplemented initially by a levy on oil consumption of roughly 0.4 US cents per litre. A pollution tax, when introduced, would become a major additional source of revenue for this fund, which would then be used to assist small-and medium-sized industries with their investments in pollution-control technologies.

There are no schemes for marketable permits in operation in Thailand.

## 2.2 USING ENVIRONMENTAL POLICY INSTRUMENTS: SOME OBSERVATIONS

### 2.2.1 Regulatory Systems versus Economic Instruments: The Policy Perspective

CAC systems of environmental regulations remain the mainstay of environmental policies for resource protection in many countries and there is considerable resistance to attempts to introduce economic policy instruments for pollution control (Kroeller, 1993; Panayotou, 1993). And, even in developed countries, fees are set too low to deter violators, and also get progressively eroded by inflation. Yet the effectiveness of CAC systems is also limited, especially in developing countries. Even with the attractiveness of economic instruments from the point of view of effective pollution control, there are several reasons why environmental regulatory systems remain the popular policy choice of governments. This subsection deals with these issues in greater detail.

*Regulatory Systems in Developing Countries*

Though CAC systems are widely used in developed and developing countries, these systems are not as effective in developing countries, for the following reasons (Panayotou, 1993).

- Litigation is not very common in many developing countries. Courts are usually used as a last resort and hence are rarely used. This being common knowledge, regulations tend to remain 'paper tigers'.
- There is also a circular logic to stricter environmental standards; given non-enforcement of existing standards, additional regulation is introduced, to which industry acquiesces, knowing that even these are not likely to be enforced.
- There is the additional problem of monitoring effluent discharges in the case of a large number of scattered small-scale industries, which, together, usually account for the bulk of the pollution.
- There is also a mismatch between the high costs of regulation, monitoring and enforcement on the one hand, and budgetary, administrative and manpower constraints on the other.
- Overcentralization of the regulatory system, which is a feature of most developing countries, also raises the costs of control.

• Rent-seeking behaviour also reduces the effectiveness of regulatory systems. Violators find it worthwhile to pay a fraction of stipulated fines as a bribe to the underpaid officials. Thus, ironically, more vigorous enforcement or stricter environmental standards mean more bribes instead of more environmental protection.

*The Case for Economic Instruments*

For one, market-based instruments can achieve the desired effect with least possible costs—both of enforcement and compliance—which is a vital consideration to developing countries with limited resources to combat environmental pollution (Panayotou, 1993).

Secondly, economic instruments present fewer opportunities for rent-seeking behaviour than regulations, and are thus likely to be more effective and more equitable (*ibid.*).

Thirdly, economic instruments are capable of generating revenues, which may be used to strengthen the enforcement administration of countries facing tight budgets and budgetary deficits (*ibid.*).

Ironically, however, CAC systems continue to be the more popular choice with implementing governments, and there are several reasons for this.

*Government Preference for Regulatory Systems*

CAC systems offer a way of hiding the true costs of environmental protection and thus help avoid distributional conflicts (Kroeller, 1993). Apart from costs of monitoring and enforcing the regulations, industries are faced with costs of complying with the regulations. These costs are substantially higher than those for complying with market-based instruments and are less obvious to the taxpayer and the consumer and are thus hidden from the public eye (Panayotou, 1993).

Also, CAC systems are a source of power and influence (often through 'personal involvement') for bureaucrats in charge of administering such systems (Kroeller, 1993).

Further, the suppression of subsidies is politically less acceptable to interest groups like trade unions because of the potential loss of jobs and social benefits (*ibid.*).

In addition, regulations are often a sure and predictable way of environmental protection, given the uncertainties of pollution taxation and the limited applicability of marketable permit systems (*ibid.*).

In contrast, there is usually a lack of understanding on how economic instruments work and how they impact growth and income distribution. Often they are dismissed as merely ways to reward polluters with the right to pollute, or to appropriate poor peoples' resources (*ibid.*).

Another concern is that by raising the costs of production, local industries will become less competitive in the international markets. Or, consumers will be faced with higher prices for consumer items in the domestic market. Either of these could be seen as an unpopular government choice and governments could feel their political stability threatened by the consequences (*ibid.*).

Such political concerns could not only delay the introduction of economic instruments into environmental policy but could also hamper their effective implementation once they are introduced. Reluctance and tentativeness in setting taxes could result in tax rates being set too low to change polluting behaviour even in the long-term (*ibid.*). Or, by the time economic incentives begin to work, a combination of inflation and political manipulation can serve to emasculate their effectiveness (*ibid.*). But even with political commitment and capability (the so-called 'political will'), the actual impact and effectiveness of economic instruments in environmental policy depends on a variety of factors and it is important for the implementing government to be aware of the institutional context within which such policies are to be introduced.

### 2.2.2 Framework for the Effective Use of Economic Instruments[7]

According to Panayotou (1993), there are five principles of effective (and politically acceptable) introduction of economic instruments:

1. setting a specific, monitorable and verifiable *ambient environmental quality target*, as the desired aim of the environmental policy;
2. trying to achieve the stated aim at the *lowest possible cost to the economy*, which includes both the enforcement and monitoring

[7] These are observations that have been made in the general context of economic instruments in environmental policy, but which have been adapted to the industrial water pollution context here.

costs to the government, as well as the costs to industry of reducing output and pollution;

3. choosing a *policy instrument which is self-financed and equitable*, with the payment being collected from the producer, who then shares it (equitably) with the consumer (say, on the basis of the price elasticity of demand for the product);

4. choosing the instrument such that any *reduction in overall industrial competitiveness is minimized;* and

5. organizing the *transition to economic instruments* in such a way that allowances are made for the adjustment costs of the interim period.

Some of these may be elaborated as follows. Given a government committed to the use of economic instruments in environmental policy, the effectiveness of such a policy will depend on the extent to which the following conditions are met (EPAT, 1993):

- *adequate technical knowledge* on the part of both the regulators and the regulated;
- a *good legal structure*, including a well-defined set of property rights;
- *competitive markets*, so that market distortions do not weaken the impact of economic instruments; and
- *adequate financial and administrative capability* with the implementing government.

In addition, to be effective, the economic instruments themselves ought to be

- *effective*—in the sense that they achieve set targets;
- *efficient*—in the sense that they achieve given targets at the least cost;
- *equitable*—in that the distribution of costs and benefits from the use of the economic instrument is perceived to be equitable (e.g. the 'polluter pays' principle); and
- *flexible*—in the sense that the economic instrument must be adaptable to changes in markets, technological knowledge and in social, political and environmental conditions (for instance, changes in the rate of inflation, or technology and hence costs of waste treatment) (*ibid.*).

Managing the transition from regulations to economic incentives can also be done so as to minimize political opposition and to maximize the effectiveness of the policy. The primary consideration ought to be a gradualised approach, rather than a one-shot imposition, since such a graduated policy:

- might make it easier for the consumer, the producers and the bureaucracy to accept the idea of economic instruments;
- allows time for adjustment, by the enforcing administration and industry, which would minimize resistance and costs of the policy; and
- would give the correct price signals to investors and innovators, for the development and use of more efficient waste control technology (Panayotou, 1993).

Three stages in the phased introduction of economic instruments may improve the chances of administrative and financial success (EPAT, 1993):

Stage 1: *Impose product charges to raise revenues.* Product charges are easier to manage, in terms of monitoring and enforcement costs, than effluent charges, and are applicable on a wider scale than marketable permits. They also have the advantage of raising revenues (unlike subsidies).

Stage 2: *Use revenues to increase technical knowledge and institutional capacity.* Such financial self-sufficiency or sustainability is especially useful in countries with tight budgets and limited administrative capacities.

Stage 3: *Use accumulating knowledge and experience to consider more effective systems.* After the requisite information base and enforcement capability has been built up, the product charge system can be supplemented with a system of effluent taxes or marketable permits, or various combinations of the two.

Despite the multiple virtues of economic instruments outlined earlier, direct regulations are recommended to control especially threatening sources of pollution, and also to provide a base framework onto which economic instruments can be grafted (*ibid.*). In addition, however, it may be useful to consider suasive instruments like voluntary pollution agreements and build up a multi-pronged environmental policy.

## 2.3 CONCLUSIONS

The experience of countries which have enforced environmental regulations to control problems of industrial water pollution underscore the dictum—to each his own. In other words, the specific context in which environmental policy has to be formulated and implemented affects not only the choice of policy instrument, but also the subsequent performance of the policy instrument in tackling the target problem of pollution. Thus, while CAC or direct regulatory systems have been decried, especially in economic theory, for being cost-inefficient, there are strong reasons that explain why governments prefer such systems, why CAC systems are the mainstay of the environmental policies of almost all countries with environmental legislation, and why Japan has had such success with CAC systems.

Marketable permits, on the other hand, have a much more limited applicability and the performance of the schemes (for water pollution control) currently in operation in the United States is not particularly encouraging.

Effluent taxes or charges appear in the environmental policies of many countries, but they are set rather low and appear to fulfil a revenue-raising role rather than one of deterring pollution by raising costs of non-compliance. Subsidies are far more prevalent, despite the negative logic of taxpayers (the victims) paying polluters through government subsidies. Hybrid instruments like taxes-and-standards, or the dual tariff scheme, or the taxes-and-rebate schemes are few and far between, despite the excellent theoretical and practical advantages that these economic instruments hold over 'pure' instruments used singly.

Even if a government has sufficient 'political will' to switch over to economic instruments from a CAC system or to introduce economic instruments into an existing regulatory framework, it may be wiser, politically and economically, to proceed in graduated steps rather than to make them a one-shot gamble. Economic instruments have tremendous potential. However, they must be chosen carefully to suit the given context, with due attention to obtaining a correct policy mix. Like any good recipe, the best results are achieved only when the ingredients are chosen and blended with care, and the dish itself is prepared by following the instructions, one step at a time.

# 3. Water Pollution Abatement: A Taxes-Standards Approach for Indian Industry[1]

A. J. JAMES and M. N. MURTY

Indian economic development since independence has been at the expense of the natural resource base that sustains such economic activity. Only in the last twenty-odd years of this forty-year period has the government taken active steps to protect the natural environment from the effects of rapid economic expansion. These measures are largely regulatory and promotionary legislation, starting from the Wildlife Protection Act of 1972 and the Water (Prevention and Control of Pollution) Act of 1974 and culminating in the Environment (Protection) Act of 1986, the amended Motor Vehicles Act of 1988 and the Public Liability Insurance Act of 1991. This emphasis on regulatory CAC measures, however, is now shifting towards the design of economic instruments for environmental protection. This chapter is an effort in the latter direction, in the particular context of industrial water pollution. The approach taken here, however, is a hybrid between the pure Pigouvian tax on effluents and the standards-based permits approach. Although other studies have been conducted in this area (e.g. Mehta, Mundle and Sankar, 1993), this analysis is broader in scope, seeking to design pollutant-specific taxes for the realization of standards laid down for the seventeen categories of highly-polluting industries identified by the Central Pollution Control Board of India. Although refinement and development is possible, this is a preliminary analysis, based on survey data, on the costs of pollution abatement in industries sampled from ten major states within the country.

[1] This chapter is mainly drawn from James and Murty (1996).

The chapter begins with a brief survey of existing policies for environmental regulation in India (Section 3.1). Section 3.2 presents the theoretical basis of the analysis, while Section 3.3 discusses the data and the methodology used to derive the pollutant-specific tax rates. The results of the analysis are presented in Section 3.4, detailed in terms of the various tax rates. It also discusses some of the cost and revenue implications of these tax rates. A concluding section draws together the various threads of the analysis.

## 3.1 THE FRAMEWORK FOR ENVIRONMENTAL REGULATION IN INDIA

The details of the various Acts of Parliament instituted to protect the environment from the effects of economic activity have been well documented elsewhere (see, *inter alia*, Murty, 1995 and Mehta, Mundle and Sankar, 1993). Suffice it here to chronologically list these legislations and to categorize them according to their objectives.

Following the Wildlife Protection Act of 1974, and the Water (Prevention and Control of Pollution) Act of 1974, the government instituted the Water Cess Act in 1977, the Forest Conservation Act in 1980, and the Air (Prevention and Control of Pollution) Act in 1981, before the Environment (Protection) Act of 1986. In 1988, the Water Cess Act of 1977 was amended as the Water (Prevention and Control of Pollution) Cess Act in 1988, along with the Air Act of 1981.[2] The Motor Vehicle Act of 1938 was also amended in 1988, keeping in mind the impact of vehicular traffic on the environment. The most recent legislation for environmental regulation is the Public Liability Insurance Act of 1991.

Of these, the Act that directly concern industrial production in India are the Water Act (1974), the Water Cess Act (1977 and 1988), the Air Act (1981 and 1988) and the Environment (Protection) Act of EPA (1986). While the first two are foundational legislations in the context of air and water pollution in the country, the last is designed to fill the gaps still remaining in the legal framework for the control of industrial pollution. The third is more a revenue-

[2] The Water (Prevention and Control of Pollution) Act and the Air (Prevention and Control of Pollution) Act are hereafter referred to as the Water Act and the Air Act, respectively.

generating legislation than a measure to restrict the consumption of water by industrial units in the interests of resource conservation.

Three issues of substance are dealt with by the former group of three Acts. First, they detail the institutional foundations of environmental regulation of industrial pollution. The early Acts provided for the setting up of Pollution Control Boards at the central and state levels, empowered to prevent, control and abate air and water pollution, and to advise governments on matters pertaining to such pollution. The Central Board was to coordinate the activities of the State Boards. The Acts also specify that industrial units are to provide on demand all information regarding their effluent and treatment methods.

Second, they regulate national standards for air and water quality, although the actual standards are laid down by the Central Pollution Control Board (e.g. CPCB, 1995), in consultation with institutions such as the National Environmental Engineering Research Institute (NEERI) in Nagpur. The standards, termed MINAS (Minimum National Standards) are defined for the country as a whole, although individual states have the option of making them more stringent. Standards are currently specified in concentration units (e.g. milligrams per litre) of certain parameters like biological-oxygen demand (BOD), chemical-oxygen demand (COD), suspended solids (SS), total dissolved solids (TDS) and pH values, in the case of wastewater and suspended particulate matter (SPM), sulphur dioxide ($SO_2$) and Nitrogen Dioxide ($NO_2$), in the case of air pollution.[3] These standards are further specified, in the case of water pollution, according to the sink for the wastewater and thus specific standards exist for discharge into surface water, streams and rivers and coastal areas.

Third, they lay down regulatory procedures for ensuring that standards are met by individual units. Following the (yearly) application by an industrial unit to the State Board, a consent order is issued once the Board is satisfied that effluent treatment will ensure the realization of the relevant water or air quality standards. But in case of non-compliance with these standards, the Boards are empowered to take action against such industrial units, with measures ranging from fines to imprisonment, to cutting-off of electricity

[3] There is a move to change over to a system of load-based standards (where load equals concentration times volume), but these have not yet been detailed.

and water connections. In addition, the EPA of 1986 gives powers to the Central government to pursue any means deemed necessary to protect and improve the quality of the environment, including co-ordinating the activities of the various state governments.

While consent orders are now commonplace bureaucratic procedure for industry and while several thousand industrial units have been taken to court for non-compliance with national or state standards for air and water quality, it is unfortunately true that there is not a single economic instrument for environmental regulation in the country. With the growing evidence that such measures are not only more effective, but achieve a given level of compliance with a lower cost to society, this is a lacuna that only adds to the social cost of pollution[4]. To see this, however, a more formal exposition is necessary.

## 3.2 THE TAXES-AND-STANDARDS APPROACH

Consider a situation where standards for ambient water quality have been laid down in concentration units (e.g. milligrams per litre) for parameters like BOD, COD, SS and pH. A pure Pigouvian tax on effluents works on the principle that the individual unit, faced with the prospect of either paying a tax or spending money on cleaning up its effluent, decides the extent of clean-up it will undertake after equating the tax rate per unit of effluent to the cost per unit of effluent (or pollution) abatement. If, however, a standard has been specified, then the tax rate is simply fixed at that level which, given the marginal abatement cost of firms, will realize the given standards. The task then, is to determine the marginal abatement cost function, at either the level of the individual firm or the industry as a whole (or indeed the country as a whole), and then to calculate the tax rate per unit of effluent necessary to realize the given standard.

In Fig. 3.1 below, if the marginal abatement cost function is given by the curve MCA and the standard is given by the distance OS, then,

[4] While the classic treatment of economic instruments for environmental regulation remains Baumol and Oates (1988), other noteworthy sources include Gunnar and Emmanuel (1991), Cropper and Oates (1992) and Eskeland and Jimenez (1992). Contrary cases are those of Japan and Korea, which used command-and-control instruments to reduce air and water pollution effectively and rapidly (O'Connor, 1992, p. 34).

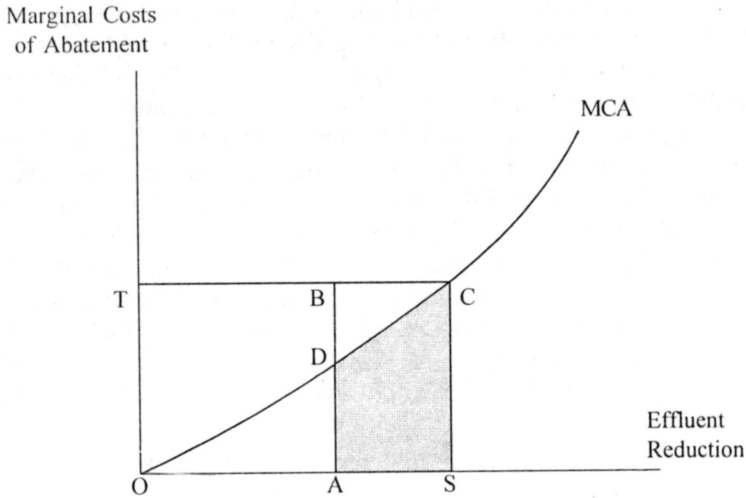


Fig. 3.1 The Taxes-and-Standards Approach

the tax rate necessary to realize the standard is OT. Both the standard and the effluent are taken to be parameter-specific, that is, in terms of BOD, COD, SS etc. Given the standard OS, the relevant tax rate (per unit of effluent) is given by OT. Thus, a firm which abates only to OA (< OS) exceeds the standards by AS, and will have to pay a total tax equal to AS times OT (the area of the rectangle ABCS thus formed).

However, this is not quite prudent economically, since the firm spends less on abatement (the shaded area ADCS) than on the tax. An economically efficient firm will thus abate till the standard OS is attained, rather than pay the tax. Somewhat obviously, the firm has no incentive to abate beyond OS since the costs of abatement are now higher than the tax that it would have to pay on those units of effluent.

The task, then, is to estimate the marginal abatement cost function and then to derive the tax rates necessary to achieve the given standards for air and water quality. There is, however, an important point that needs to be noted in this connection. That is, besides the pure efficiency gains of an economic instrument over CAC type of regulation (i.e. the same improvement in environmental quality can be achieved at a lower social cost), there is another hidden advantage that an economic instrument has. As Pezzey (1994) points out, the marginal abatement cost incurred by firms in a regulated environment will tend to fall once an economic policy instrument is in place. Either through moves to adopt more efficient clean-up technology, or through more efficient use of existing equipment, the marginal costs of abatement will decrease with a move to a more efficient market-based control mechanism (and not just to a *fully* efficient market mechanism). Given that tax rates would fall accordingly, there is, thus, a hidden incentive for firms to support a market-based instrument for pollution control rather than a mere regulatory mechanism.

In Fig. 3.2 below, the efficiency gain from moving from a regulated policy environment to a market-based policy regime is the decrease in the industry's marginal costs of abatement from MCA I to MCA II. This reduces the marginal tax rate for a given excess discharge of effluent (say, OA) from OT to OW and, correspondingly, the tax liability shrinks to the shaded area AEFS.

But this is necessarily a two-step procedure, with the initial level of tax rate OT providing the incentive to firms to increase the efficiency of their abatement activities, thus moving down to a lower marginal abatement cost schedule, and then a secondary assessment of these costs yielding the new lower tax rate OW. It must, however,

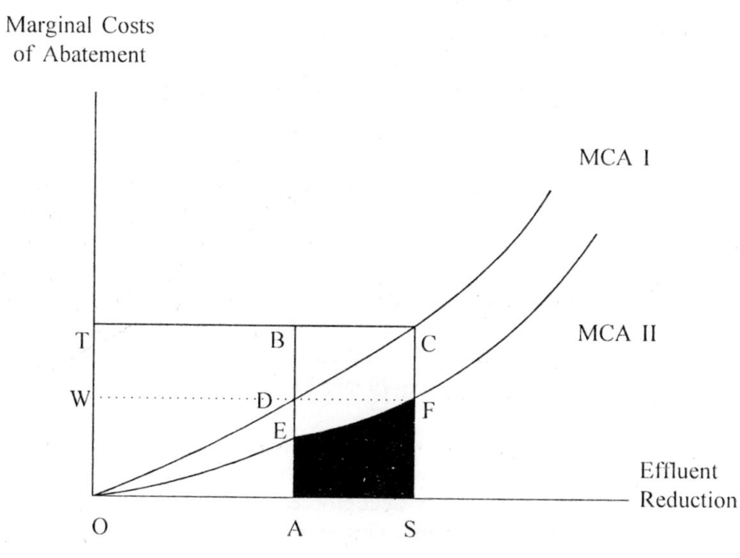


Fig. 3.2 Efficiency Gains with a Market-based Policy Regime

be said that the taxes-and-standards approach is viable only for large- and medium-scale industrial units in India, since the small-scale units are usually prevented by diseconomies of scale from investing in their own effluent-treatment plants.[5] The mechanics of the procedure to derive the pollution tax rate OT, however, is perhaps clearer with some formalization.

### 3.2.3 Total and Marginal Abatement Costs and the Tax Rate

Assuming that there are only two inputs into the effluent treatment plant, labour and capital, the general form of the (total) abatement cost function to be estimated is:

$$C = f\left(Volume, \frac{BODEFF}{BODINF}, \frac{CODEFF}{CODINF}, \frac{SSEFF}{SSINF}, w_k, w_L\right)$$

where C = the total annualized cost of the effluent treatment plant;
VOLUME = the annual volume of wastewater treated by the plant;
BODINF = concentration of BOD in the influent[6] (milligrams per litre);
BODEFF = concentration of BOD in the effluent (milligrams per litre);
CODINF = concentration of COD in the influent (milligrams per litre);

[5] In standard economic parlance, this is an example of a non-convex technology (with a falling MCA), in which case the usual logic of a pollution tax will not work. In the small-scale case, however, the solution proposed and implemented in several parts of the country are CETPs, which overcomes the scale problems of individual units and provide a cost-effective service for such industrial units. The problems of organizing CETPs, of sharing the costs and the responsibilities of operation, in the Indian context, have been documented elsewhere (see, Misra and Murty, 1995a; Murty, 1995b; Murty and Prasad, 1995).

The marginal cost, in other words, is the cost of reducing the COD load (concentration times volume) by 1 milligram.

[6] 'Influent' refers to the wastewater proceeding out of the production plant, prior to treatment, while 'effluent' refers to wastewater after treatment, prior to discharge into the environment.

CODEFF = concentration of COD in the effluent (milligrams per litre);

SSINF = concentration of SS in the influent (milligrams per litre);

SSEFF = concentration of SS in the effluent (milligrams per litre);

$w_K$ = the price of the capital used in the effluent treatment plant; and

$w_L$ = the price of the labour used in the effluent treatment plant.

Assuming a linear form for the cost function, it can be written in logarithmic form as

$$\log C = \log VOLUME + \log \left[\frac{CODEFF}{CODINF}\right] + \log \left[\frac{BODEFF}{BODINF}\right] + \log \left[\frac{SSEFF}{SSINF}\right] + \log w_K + \log w_L$$

Correspondingly, the equation to be estimated econometrically is

$$\log C = \alpha_0 + \alpha_V \log VOLUME + \alpha_C \log \left[\frac{CODEFF}{CODINF}\right] + \alpha_B \log \left[\frac{BODEFF}{BODINF}\right] + \alpha_S \log \left[\frac{SSEFF}{SSINF}\right] + \log w_K + \log w_L$$

Four marginal abatement cost functions can be derived: one each with respect to the three parameters BOD, COD and SS, and one with respect to the volume of wastewater treated. The marginal abatement cost functions with respect to COD and the volume of wastewater are derived explicitly below, and the others can be derived similarly.

### 3.2.4 Marginal Abatement Cost Function with respect to COD

There are three points to be noted about the marginal abatement cost function (with respect to COD in wastewater). Firstly, the cost of abatement is incurred in the context of reducing the *quantity* of COD (or BOD or SS) in the effluent. Hence, the appropriate partial derivative of the total abatement cost function is that with respect to the product of CODEFF (or BODEFF or SSEFF) and the VOLUME of wastewater in which this concentration is found.

Secondly, since this marginal cost represents the cost of reducing the quantity of BOD (or COD or SS) in the influent wastewater, and since the variable in question is the ratio of BODINF/BODEFF, it is clear that a reduction in the quantity of BOD in the effluent represents as *increase* in the ratio.

Thirdly, since these are marginal costs for the reduction in the COD (or BOD or SS) load (concentration times volume), and since the volume matters in the total costs of abatement, the marginal cost has to be divided by the (average) volume of wastewater treated.

The marginal abatement cost functions, thus, have a negative sign and are divided by the volume of wastewater treated. The marginal abatement cost function corresponding to a decrease in COD in wastewater, accordingly, is given by

$$MCA_{COD} \equiv -\frac{\partial C}{\partial(CODEFF \times VOLUME)} \equiv -\frac{\partial C}{\partial(COD\ load)}$$

$$= -\frac{\alpha_C \cdot C}{(CODEFF \times VOLUME)}$$

As is conventional, in the actual estimation, the full form of the (estimated) total cost function is used and the level of all other variables—except that with respect to which the derivation is carried out (i.e. CODEFF)—are taken to be at their mean values within the sample. However, since these can vary, the actual estimates have used maximum, mean and median values of CODINF and VOLUME. These are detailed in Section 3.4, which discusses the different estimation trials and their results.

### 3.2.5 Marginal Abatement Cost Function with respect to Volume of Wastewater

This marginal cost function is the standard partial derivative of the total abatement cost function with respect to the volume of wastewater treated (in kilolitres) per annum. If this cost is given by $MCA_{VOL}$, the function is

$$MCA_{VOL} \equiv -\frac{\partial C}{\partial VOLUME} = \frac{\alpha_V \cdot C}{VOLUME}$$

Again, although the estimated total cost function uses mean values of all variables except VOLUME, simulations have used maximum and median values of CODINF as well.

### 3.2.6 The Effluent Tax Rate

The tax rate is calculated simply by substituting the actual (average) concentration of the effluent (e.g. CODEFF) in the wastewater by the standard (i.e. by 250 milligrams per litre). The procedure can be repeated to get the tax rate required to support the BOD standard or the standard for SS (see CPCB, 1995, for a detailed account of these standards).

The methods used to collect the data necessary for the estimation and those of the estimation itself are detailed in the next section, while the results of the estimation, including the various tax rates, are detailed in the following section.

## 3.3 ESTIMATING ABATEMENT COST FUNCTIONS FOR WASTEWATER TREATMENT

### 3.3.1 The Data

The data required to estimate the abatement cost function are not collected in full by the Central Pollution Control Board or the various State Pollution Control Boards or any other government or private agency. Consequently, a questionnaire-based survey was conducted, backed-up by selected factory visits, in order to compile the data on influent and effluent quality as well as on costs and input prices. These questionnaires were sent only to large- and medium-sized firms falling under the highly polluting industries category identified by the Water Cess Act of 1977 and specified further by the Central Pollution Control Board, New Delhi.[7] A total of 131 firms responded from ten of the major industrialized states in the country, although

[7] The categories of highly-polluting industries identified by the Water Cess Act of 1977 (Schedule 1) are: (1) ferrous metallurgical industries, (2) non-ferrous metallurgical industries; (3) mining industry; (4) ore-processing industry; (5) petroleum industry; (6) petro-chemical industry; (7) chemical industry; (8) ceramic industry; (9) cement industry; (10) textile industry; (11) paper industry; (12) fertilizer industry; (13) coal (including coke) industry; (14) power (thermal and diesel) generating industry; and (15) processing of animal or

only 82 questionnaires yielded full data.[8] The sample size for the estimation of the aggregate marginal abatement cost function, therefore, is 82.

### 3.3.2 The Total Abatement Cost Function

The aggregate (total) abatement cost function to be estimated is taken to be a function of the quality of effluents and influents and of the input prices. Only two inputs are taken, capital and labour. Further, instead of taking influent and effluent quality as separate variables, the ratio of influent quality to effluent quality is taken in the cost function. This is the more intuitive approach, since costs are incurred in the process of *reducing* the amounts of BOD, COD and SS in the influent wastewater.

The first practical problem faced in computing the total abatement cost of various firms is that of computing the capital cost. Since the data in the filled-in questionnaire refers to capital stock of different vintages, they were first converted into constant 1993–4 values. Next, the price of capital was estimated (see below) and used to convert these into annual estimates of the cost of capital.

A related problem is that of calculating the price of capital. This was derived by a value-added approach and using secondary data from the Annual Survey of Industries, assuming that the capital and labour invested in the effluent treatment plant are perfectly substitutable with those employed in the main plant. The value-added by the production plant is the turnover less the sum of material inputs used in production. This is also equivalent to the value added by the primary factors, labour and capital. Subtracting labour costs yields the value-added due to capital alone. Dividing this by the relevant estimate of invested capital (available from the data of the Annual Survey of Industries), yields the rate of return on capital. Adding to this the rate of depreciation yields an estimate of the price of capital services. Since the (financial) capital used in the effluent treatment plant is assumed to be shifted out of main plant to this alternative

---

vegetable products industry. The Central Pollution Control Board has further specified seventeen categories of highly-polluting industries within these general categories.

[8] The states are Tamil Nadu, Kerala, Andhra Pradesh, Karnataka, Maharashtra, Gujarat, Orissa, West Bengal, Delhi and Haryana.

use, the same price applies to the capital invested in the effluent treatment plant.[9]

## 3.4 ESTIMATED TOTAL AND MARGINAL COSTS OF ABATEMENT AND TAX RATES

There is one additional feature of the estimated (as opposed to theoretically derived) marginal costs of abatement that deserves prior mention: the question of scale. Since the influents and effluents are given in concentration units (i.e. milligrams per litre) and since the volume of wastewater is given in kilolitres (i.e.1000 litres), it is appropriate to convert the concentrations also into milligrams per kilolitres. This is done simply by multiplying the concentrations by 1000. Further, since the cost is in terms of millions of rupees per annum and it is desirable to have the tax per unit of effluent in terms of rupees, the marginal cost is multiplied by a factor of 1,000,000. Finally, since the tax on annual wastewater quality is now in terms of rupees per milligram, and the excess is likely to be in terms of even kilograms (given the high volumes treated per plant), it is sensible to have a tax at least in terms of rupees per 100 grams, by multiplying the marginal cost estimate by 100 × 1000.[10]

### 3.4.1 Estimated Total Costs of Abatement

Although a full translog specification, a truncated translog specification and a full Cobb-Douglas specification of the total abatement cost function were tried, these were found to give a poor fit to the data (see Appendices A 3.1 and A 3.2). This hinted at a possible multicollinearity between the three ratios, which was confirmed by a high correlation between pairs of effluent ratios and by a rise in the $R^2$ (and adjusted $R^2$) and in the significance of the ratios when tried singly. Since the $R^2$ value and the significance levels were the

[9] Thus, Total Value Added = Turnover–Material Cost
= Value Added by Capital and Labour (i.e. the Primary Factors).
Value Added by Capital = Total Value Added–Value Added by
= Labour (i.e. labour cost).
The Rate of Return on Capital = Value Added by Capital/Invested Capital
The Price of Capital = Rate of Return on Capital + Rate of Depreciation.

[10] 1000 milligrams make a gram.

highest for the specification with the COD ratio and the input prices, this was chosen as the base estimation. In addition, the cost function was estimated with (1) the COD ratio with and without input prices; (2) the BOD ratio without input prices, and (3) an index which converted BOD ratios into COD ratios using fixed weights.[11] The four corresponding regression equations are detailed below while the results of these estimations are presented in Table 3.1.

*Estimation 1*

$$\log C = \alpha_0 + \alpha_V \log VOLUME + \alpha_C \log \left[\frac{CODINF}{CODEFF}\right] + \alpha_{WK} \log w_K + \alpha_{WL} \log w_L$$

*Estimation 2*

$$\log C = \alpha_0 + \alpha_V \log VOLUME + \alpha_C \log \left[\frac{CODINF}{CODEFF}\right]$$

*Estimation 3*

$$\log C = \alpha_0 + \alpha_V \log VOLUME + \alpha_B \log \left[\frac{BODINF}{BODEFF}\right]$$

*Estimation 4*

$$\log C = \alpha_0 + \alpha_V \log VOLUME + \alpha_C \log \left[\frac{CODININF}{CODINEFF}\right]$$

where CODININF is the COD index of influent quality and CODINEFF, the COD index of effluent quality.

Table 3.1
Parameter Estimates of Total Abatement Cost Functions 1 to 4

| | Estimation 1 | Estimation 2 | Estimation 3 | Estimation 4 |
|---|---|---|---|---|
| $\alpha_O$ | –1.06 | –5.05*** | –4.87*** | –5.03*** |
| $\alpha_V$ | –0.12** | –0.20 | –0.47*** | –0.48*** |
| $\alpha_C$ | 0.46*** | 0.49 | – | –0.21*** |
| $\alpha_B$ | – | – | –0.18*** | – |
| $\alpha_{WK}$ | 1.16*** | – | – | – |
| $\alpha_{WL}$ | 0.53*** | – | – | – |
| $R^2$ | 0.75 | 0.57 | 0.57 | 0.57 |
| Adjusted $R^2$ | 0.74 | 0.56 | 0.56 | 0.56 |

*Notes*: All t-ratios used are heteroscedasticity-corrected t-statistics, ** denotes significance at the 5 per cent level, *** denotes significance at the 1 per cent level.

[11] These indexes were constructed as follows. First, the co-efficient of the ratio of BODINF/BODEFF in the regression of the ratio of CODINF/CODEFF

### 3.4.2 Estimated Pollutant-specific Marginal Costs of Abatement and Tax Rates

Equation 1 was selected to base marginal cost calculations because of the better fit of this specification to the data (the higher adjusted $R^2$). While marginal cost calculations from these estimates take average values for the other variables in total cost, two variations were introduced. First, the volume of wastewater was allowed to vary in order to determine the extent of economies of scale. The maximum volume reported in the sample was used as one extreme, but since the minimum volume was extremely low, the median value was chosen. But even the median value is rather small in comparison to the mean, and thus the marginal abatement cost is substantially higher with median volume of wastewater. This confirms the presence of extensive economies of scale.

Second, the influent quality was allowed to vary. The maximum and the median levels were chosen once again, and marginal costs were observed to rise with a fall in influent quality. This was however done only in the case of the estimates from the regression with COD ratio, volume and the input prices as independent variables.

Since the variation is reflected in the tax rates computed in each case, and since the tax rates are the variables of interest, the results are presented in terms of tax rates.

Table 3.2 shows a sharp increase in tax rates as wastewater volume decreases, indicating that there are significant economies of scale in pollution abatement technologies. The graphs of the three marginal abatement cost curves for any given level of CODINF (see Fig. A 3.3.1, Appendix A 3.3) show that the marginal cost curve corresponding to the median level of wastewater volume is far higher than that for the average level of wastewater volume, which in turn is higher than that corresponding to the maximum level of wastewater volume. The Appendix graphs (Fig. A 3.3.2) also show that as the influent concentration of the pollutant (COD) increases, for any given level of wastewater, the marginal costs of treatment rises as well. The tax rates in Table 3.3 reflect this trend.

---

on the former ratio was obtained (say, ß) used to construct weights. While the COD ratio of any given firm was assigned a weight of 1/(1 + ß), the BOD ratio was assigned a weight of ß/(1+ß). The resulting weighted index is the independent variable CODININF/CODINEFF.

Table 3.2
Tax Rates in Rupees for 100 Grams Reduction in COD Per Annum: Estimation 1

| | Maximum VOLUME | Average VOLUME | Median VOLUME |
|---|---|---|---|
| Maximum CODINF | 0.10 | 0.41 | 8.08 |
| Median CODINF | 0.06 | 0.24 | 4.82 |

*Notes:* 1. Maximum annual volume (in kilolitres) is 47,320,000, while the average and median volumes are 3,676,400 and 12,570 respectively.
2. Maximum CODINF (in milligrams per litre) is 1,06,000, and median, 1,340.

With equation 2 (with COD ratio and wastewater volume as the independent variables) the pattern of tax rates is the same, with the rate rising as the volume of wastewater decreases. A similar pattern characterizes the estimations with BOD ratio and the COD index replacing COD ratio among the independent variables (estimations 3 and 4).

Table 3.3
Tax Rates in Rupees for 100 Grams Reduction in COD Per Annum: Estimations 2, 3 and 4

| | Average CODINF | Average BODINF | Average COD Index |
|---|---|---|---|
| Maximum Volume | 0.15 | 1.08 | 0.13 |
| Median Volume | 8.97 | 77.42 | 8.67 |

*Note:* The first and third columns of tax rates refer to 100 grams reduction in COD per annum, while the second column refers to 100 grams reduction in BOD per annum.

These tax rates, by themselves, do not give any indication of the relative magnitudes of cost to the concerned firm. For the firms in the sample with excess COD concentration in their wastewater, the tax liabilities calculated using the average tax rate of Rs 0.32 per 100 grams are given in Table 3.4, along with the total tax liability as a percentage of its annual turnover.

There are two points that emerge from this exercise. First, only 21 out of the 82 firms in the sample exceed the COD standard. That is, less than a third of the firms in the sample are defaulters. Of these, about 10 firms have excess concentrations of COD equal to or less than 150 milligrams per litre. About 8 firms have an excess COD concentration of between 150 and 1000 milligrams per litre, while 3 firms have substantial excesses over the standard (over 6000 milligrams per litre). However, their tax liability is based on excess concentration times the volume of wastewater, and hence a firm with only 84 milligrams per litre excess COD and 6.39 million kilolitres of wastewater volume per annum ends up with a tax liability of over 1.6 million rupees. Further, the maximum tax liability of Rs 23.5 million (Rs 2.35 crore) rupees falls on a firm with an excess COD concentration of 825 milligrams per litre but over 9 million kilolitres of wastewater volume per annum, and a firm with an excess of nearly 27,000 milligrams per litre in its effluent is liable to pay only around Rs 8.8 million (Rs 88 lakh) since its volume of wastewater per year is just 0.1 million kilolitres.[12]

Table 3.4
Cost Implications of the Effluent Tax: Liability as a Proportion of Turnover

| Firm | Excess COD concentration (milligrams per litre) | Annual volume of wastewater (kilolitres) | Annual Tax Liability (Rupees) | Annual Turnover (Rupees millions) | Tax liability as a percentage of turnover (%) |
|---|---|---|---|---|---|
| 1 | 6 | 67,450 | 1,273.02 | 235.93 | 0.0005 |
| 2 | 30 | 30,175 | 2,847.55 | 152.01 | 0.0019 |
| 3 | 38 | 74,550 | 8,911.17 | 122.40 | 0.0073 |
| 4 | 65 | 106,500 | 21,775.40 | 223.93 | 0.0097 |
| 5 | 68 | 793,070 | 169,638,00 | 999.50 | 0.0017 |
| 6 | 72 | 29,465 | 6,673.33 | 26.78 | 0.0025 |
| 7 | 75 | 639,000 | 150,753.00 | 3,290.00 | 0.0046 |
| 8 | 84 | 6,390,000 | 1,688,430.00 | 1,400.00 | 0.1206 |
| 9 | 90 | 106,500 | 30,150.60 | 815.38 | 0.0037 |
| 10 | 150 | 42,600 | 20,100.40 | 150.00 | 0.0134 |
| 11 | 230 | 6,035 | 4,366.25 | 23.14 | 0.0188 |
| 12 | 240 | 2,564,880 | 1,936,340.00 | 180.35 | 1.0737 |
| 13 | 350 | 514,750 | 5,66,719.00 | 1,078.90 | 0.0525 |
| 14 | 500 | 2,662,500 | 4,187,580.00 | 407.75 | 1.0270 |
| 15 | 650 | 2,662,500 | 5,443,850.00 | 184.00 | 2.9586 |
| 16 | 825 | 9,052,500 | 23,492,300.00 | 7,279.50 | 0.3183 |
| 17 | 950 | 284,000 | 848,683.00 | 180.05 | 0.4714 |
| 18 | 1,085 | 7,100 | 24,232.10 | 60.00 | 0.0439 |
| 19 | 6,250 | 79,875 | 1,570,340.00 | 10.31 | 15.2313 |
| 20 | 26,924 | 104,370 | 8,839,320.00 | 39.30 | 22.4919 |
| 21 | 32,250 | 213,000 | 21,607,900.00 | 146.70 | 14.7293 |

[12] Note however that in the case of an economic instrument like the pollution tax, polluting firms informed of excess emission, would prefer to spend on

Secondly, the ratio of annual tax to turnover becomes significant (i.e. above 4 per cent) for only 4 firms out of 21 (or about 14 per cent of the defaulters). While it is true that firm profits are a better indicator of the 'bite' of the tax, such a ratio gives a rough relative picture of the effectiveness of the tax.

Finally, in terms of revenue to the government, it is illustrative that from this handful of firms, the revenue raised (the sum of the tax liabilities) is around Rs 68 million. There is thus considerable potential in such a tax for enforcing abatement and generating revenue.

### 3.4.3 Marginal Costs with respect to Volume of Wastewater

The graphs in Appendix A 3.3 show that as the quality of the influent falls, that is the COD concentration in the influent rises, the marginal cost with respect to the volume of wastewater rises.

## 3.5 CONCLUSIONS

The basic conclusion is that a market-based hybrid instrument like a taxes-and-standards scheme can have significant incentives for industrial pollution control in India. In contrast to command and control measures or regulations, such an approach can benefit not only those affected by pollution and the government but also, the polluting industries themselves.

For these industries, in addition to the positive economic incentive of having tax rates that are at least as high as the cost of effluent treatment, there is also a secondary incentive in the form of lower tax rates in the next round of tax revisions. This is because as firms adjust to the new environment of a market-based policy instrument like the pollution tax, their efficiency in pollution abatement would tend to rise, thus bringing down the costs of effluent treatment. Since the tax is based on the costs of abatement, this would, in turn, bring down the tax rates.

Finally, such a policy may lead to a rise in supply of improved and multipurpose abatement techniques aimed at recycling and reusing material from the wastewater. This would then add another

---

effluent control rather than on paying the tax. Given time, therefore, the total taxes paid could be far short of the assessed tax liability.

incentive to firms for installing effective equipment to control water pollution.

This analysis showed, in the context of a selected number of firms from all over the country, the methodology of determining a set of tax rates to enforce the existing standards for water quality in the country. Such analysis can be easily extended, provided adequate data is available, to the case of pollutant-specific or even industry-specific tax rates.

The empirical analysis revealed several problems particular to the Indian context, but also suggested ways of overcoming these problems and arriving at a set of tax rates. In particular, secondary data is unavailable and has to be generated from primary surveys. Also, capital prices for a cross-sectional econometric exercise have to be generated, and the chapter suggested (and used) a method to overcome this problem. Finally, the tax rates have to be specified in terms of pollution loads, which is not possible due to the presence of scale economies. In order to overcome this problem, while the econometric estimation is in terms of pollution loads (i.e. concentration times volume), the final tax rates are in terms of concentration units (i.e. rupees per 100 grams of BOD reduction per annum).

There is considerable scope for improvement over this preliminary analysis of the Indian case. More such studies are desperately needed so that at least a basic panel database can be built up in the future. Also, more econometric studies will enable researchers to arrive at a standard specification of the total and marginal abatement cost functions, which can be used subsequently at the policy level to set the required tax rates. Also, the delineation of region-specific and industry-specific load-based pollution standards will go a long way in ensuring that meeting pollution standards will mean a pollution-free environmental sink.

The main contribution of this chapter is to provide empirical support for a taxes-and-standards policy instrument, in the hope that it will stimulate further work in this direction, both from researchers and from policy makers.

# APPENDIX A 3

## Appendix A 3.1: Translog Specifications of the Total Abatement Cost Function

The translog functional form of the total abatement cost function is as follows:

$$LTC = \alpha_0 + \alpha_V\, LVOL + \alpha_c\, LCR + \alpha_B\, LBR + \alpha_s\, LSR + \alpha_{WK}\, LWK + \alpha_{WL}\, LWL +$$
$$\alpha_{VV}\left(\frac{LVOL^2}{2}\right) + \alpha_{VC}\, LVOL \times LCR + \alpha_{CB}\, LVOL \times LBR + \alpha_{CS}\, LVOL \times LSR +$$
$$\alpha_{CC}\left(\frac{LCR^2}{2}\right) + \alpha_{CB}\, LCR \times LBR + \alpha_{CS}\, LCR \times LSR +$$
$$\alpha_{BB}\left(\frac{LBR^2}{2}\right) + \alpha_{BS}\, LBR \times LSR +$$
$$\alpha_{SS}\left(\frac{LSR^2}{2}\right) +$$
$$\alpha_{WKK}\left(\frac{LWK^2}{2}\right) + \alpha_{LWKL}\, LWK \times LWL +$$
$$\alpha_{WLL}\left(\frac{LWL^2}{2}\right)$$

where

| | | |
|---|---|---|
| LTC | = | logarithm of total annual cost of the effluent treatment plant; |
| LVOL | = | logarithm of the total annual volume of wastewater generated; |
| LCR | = | logarithm of the ratio of COD in the treated effluent to that in the untreated influent; |
| LBR | = | logarithm of the ratio of BOD in the treated effluent to that in the untreated influent; |
| LSR | = | logarithm of the ratio of SS in the treated effluent to that in the untreated influent; |
| LWK | = | logarithm of the price of capital; and |
| LWL | = | logarithm of the price of labour. |

Further, it is assumed that the direct cost function is separable into input prices and the other components. This total abatement cost function was estimated with and without input prices. A 'truncated' version of this full translog function was also tried again, with and without input prices:

$$LTC = \alpha_0 + \alpha_V\, LVOL + \alpha_c\, LCR + \alpha_B\, LBR + \alpha_s\, LSR + \alpha_{WK}\, LWK + \alpha_{WL}\, LWL +$$
$$\alpha_{VC}\, LVOL \times LCR + \alpha_{CB}\, LVOL \times LBR + \alpha_{CS}\, LVOL \times LSR$$

The parameter estimates from these four estimations are given below. Since the parameters are not well estimated by this functional form (indicated by the low t-ratios), the Cobb-Douglas form was tried subsequently. The results of the translog specification are detailed in Table A 3.1 below.

Table A 3.1
Parameter Estimates of Total Abatement Cost

| Parameters | Full Translog with Prices | Full Translog without Prices | Truncated Translog with Prices | Truncated Translog without Prices |
|---|---|---|---|---|
| $\alpha_0$ | 0.84 | 1.75 | 0.45 | –4.74*** |
| $\alpha_v$ | 0.65 | –0.24 | 0.36*** | 0.46*** |
| $\alpha_C$ | –0.77 | 1.21 | –0.87 | –0.33 |
| $\alpha_B$ | 0.91 | 0.29 | 0.56 | –0.06 |
| $\alpha_S$ | 0.79 | 0.86 | 0.69** | 0.33 |
| $\alpha_{WK}$ | 1.93 | – | 1.19*** | – |
| $\alpha_{WL}$ | 1.16* | – | 0.59*** | – |
| $\alpha_{VV}$ | –0.02 | 0.04 | – | – |
| $\alpha_{VC}$ | 0.04 | –0.08 | 0.06 | 0.02 |
| $\alpha_{VB}$ | –0.05 | –0.01 | –0.04 | – |
| $\alpha_{VS}$ | –0.05 | –0.04 | –0.05** | –0.02 |
| $\alpha_{CC}$ | –0.02 | 0.21 | – | – |
| $\alpha_{CB}$ | 0.10 | –0.03 | – | – |
| $\alpha_{CS}$ | –0.17 | –0.06 | – | – |
| $\alpha_{BB}$ | –0.04 | 0.15 | – | – |
| $\alpha_{BS}$ | 0.06 | –0.07 | – | – |
| $\alpha_{SS}$ | 0.18** | 0.26** | – | – |
| $\alpha_{WKK}$ | 0.86 | – | – | – |
| $\alpha_{WKL}$ | –0.25 | – | – | – |
| $\alpha_{WLL}$ | 0.26** | – | – | – |
| $R^2$ | 0.80 | 0.64 | 0.76 | 0.57 |
| Adjusted $R^2$ | 0.74 | 0.56 | 0.73 | 0.53 |

*Notes:* * denotes significance at the 10 per cent level, ** denotes significance at the 5 per cent level and *** denotes significance at the 1 per cent level. Standard errors used are heteroscedasticity-corrected.

## Appendix A 3.2: Cobb-Douglas Specifications of the Total Abatement Cost Function

The following versions of the Cobb-Douglas specification of the total abatement cost function were estimated, and the results are detailed in Table A 3.2 below.

Regression 1 : LTC = f(LVOL, LCR, LBR, LSR, LWL, LWK)
Regression 2 : LTC = f(LVOL, LCR, LBR, LSR)
Regression 3 : LTC = f(LVOL, LCR, LWL, LWK)
Regression 4 : LTC = f(LVOL, LCR)
Regression 5 : LTC = f(LVOL, LBR, LWL, LWK).
Regression 6 : LTC = f(LVOL, LBR)
Regression 7 : LTC = f(LVOL, LSR, LWL, LWK)
Regression 8 : LTC = f(LVOL, LSR)

Table A 3.2
Parameter Estimates for Cobb-Douglas Total Abatement Cost Functions

| Parameters | Regression 1 | Regression 2 | Regression 3 | Regression 4 | Regression 5 | Regression 6 | Regression 7 | Regression 8 |
|---|---|---|---|---|---|---|---|---|
| $\alpha_0$ | –0.98 | –4.99*** | –1.06 | –5.05*** | –0.80 | –4.87*** | –0.42 | –4.36*** |
| $\alpha_V$ | 0.47*** | 0.49*** | 0.46*** | 0.49*** | 0.44*** | 0.47*** | 0.44*** | 0.46*** |
| $\alpha_C$ | –0.20 | –0.13 | –0.12** | –0.20** | – | – | – | – |
| $\alpha_B$ | 0.06 | –0.10 | – | – | –0.07 | –0.18** | – | – |
| $\alpha_S$ | 0.05 | 0.05 | – | – | – | – | 0.00 | –0.06 |
| $\alpha_{WK}$ | 1.18*** | – | 1.16*** | – | 1.16*** | – | 1.20*** | – |
| $\alpha_{WL}$ | 0.54*** | – | 0.53*** | – | 0.54*** | – | 0.56*** | – |
| $R^2$ | 0.75 | 0.57 | 0.75 | 0.57 | 0.74 | 0.57 | 0.74 | 0.54 |
| Adjusted $R^2$ | 0.73 | 0.55 | 0.74 | 0.56 | 0.73 | 0.56 | 0.73 | 0.53 |

*Notes:* * denotes significance at the 10 per cent level, ** denotes significance at the 5 per cent level and *** denotes significance at the 1 per cent level. Standard errors used are heteroscedasticity-corrected.

## APPENDIX A 3.3

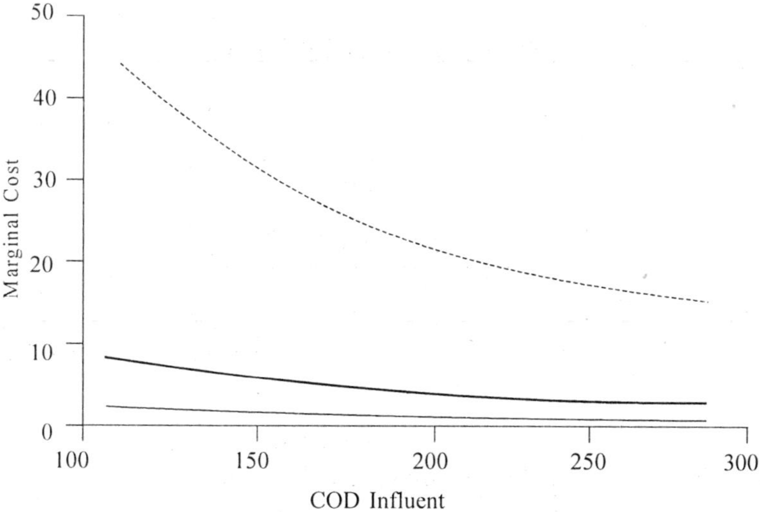


........ Median volume — Average volume
— Maximum volume

Fig. A 3.3.1 Different Wastewater Volumes for the Same (average) COD in Influent

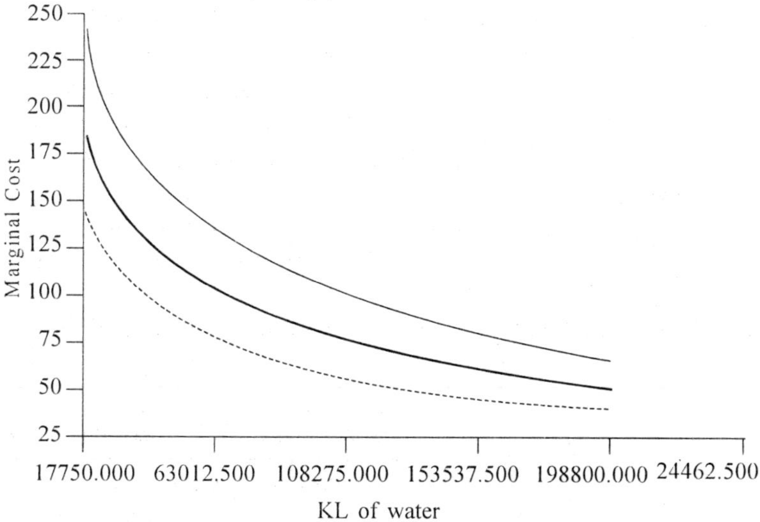


........ Maximum COD Influent — Median COD Influent
— Average COD Influent

Fig. A 3.3.2 Different Levels of COD in Influent

# PART II
# Collective Action for Pollution Regulation

# Introduction to Part II

The use of economic instruments and command controls to deal with an environmental externality hinges on the assumptions that the government is benevolent and there are supporting legal and other institutions. The government is not benevolent in most of the developing countries and this may be true even in some developed countries. In many countries, required environmental laws may be absent or less developed and even if they are present, they are ineffective due to various reasons (Pargal and Wheeler, 1996). Empirical evidence from the pollution abatement technologies used especially in controlling water pollution points out that there are non-convexities in pollution abatement (Dasgupta and Murty, 1985; Shekhar, Mundle and Sankar, 1993; James and Murty, 1996; Misra and Murty, 1995) so that one cannot be sanguine about the general application of economic instruments discussed above. It is therefore important to look for institutions alternative to government to deal with environmental externalities. The alternatives can be (a) collective action or cooperative solution involving all the agents relevant for controlling an environmental externality and (b) a purely market option. In all these alternatives, the role of the government is minimal, if not zero.

Inability to create property rights to resources which are non-exclusive and undepletable is seen as an impediment to the development of markets for the environmental externalities. Coase (1960) has argued that many types of externalities can be optimally controlled by creating property rights to relevant agents. It can be shown that an institutional setting in which there is costless bargaining between the generator and the receiver of an externality results in its optimal control, and that the final result of bargaining is invariant to the initial allocation of property rights (for example, whether the right to a clean environment is vested in the affected people or whether the right to pollute is given to polluters). As given by Coase (1988), the Coase theorem is stated as follows: With zero transaction costs, private and social costs will be equal.... and the value of production will be maximized. In this case, the government has to play only the minimal policing role of creating and protecting the initial allocation of property rights. Unlike the pollution tax, such an institutional setting is effective in controlling pollution even with non-convex abatement technology. Even with positive transaction cost, bargaining results in net welfare enhancing reductions in environmental pollution. Coasean solution provides an institutional alternative to economic and command-and-control instruments and it especially assumes importance when the quality of government is in doubt.

The key assumption in the Coasean solution is that all the externalities are captured in the value of property rights and there are incentives for gainful bargaining. It can work well with externalities on a smaller scale or local externalities of the type described in Coase's seminal paper (examples in Coase, 1960: a building that blocks windmill's air currents; a confectioner's machine that disturbs a doctor's quiet, etc.). Many environmental externalities occur on a grander scale with a large number of receivers and many times, a good number of generators (Example: pollution of a river and the atmosphere) which makes the job of creating property rights and facilitating trade onerous. One may surmount this problem by creating a common property right of an environmental resource to an all inclusive group of people affected by its degradation so that there can be incentives for gainful bargaining between a group of affected people and a group of generators *a la* Coase. Another problem for defining property rights for an environmental resource arises because it is a stock affecting the welfare of both present and future generations. Capitalization of future benefits from this resource is not possible because property rights to future generations of affected people cannot be defined. However, it may be possible to define property right to a polluting firm since it lives into perpetuity (Nalebuff, 1997). While taking a decision for using the resource, a firm can take into account the capitalized value of future benefits from the resource (Nalebuff, 1997). One approach to take care of a future generation is to consider the government as its representative. The government can compete in the market for environmental property rights of the future generation and pay for it by issuing a debt which has to be serviced by the future. Another approach is based on the assumption that the present generation has a bequest motive to the future and wants to bequeath to the future the preserved resources. However, both government intervention and bequest motive are outside the scope of Coase's property rights approach.

A closely-related problem to the property rights solution discussed above is the management of common pool resources, resources which possess both private and public good characteristics (examples: village commons, forest lands, fishing grounds etc.). In terms of property rights, these resources can be managed either under any of the property rights regimes: government property, common property and a private property or as a free access resource. Common pool resources as free access resources (resources without property rights) face a prisoner's dilemma type of problem subject to free riding (Wade, 1987). Management of these resources through government property has been taken as a solution to this problem. However, experience in both developed and developing countries shows that government property regime is not successful in the management of common pool resources like forests (Bromely and Chapagain, 1984; Chopra et.al, 1990; Saxena, 1993). Collective action of local beneficiaries resulting in a common property rights regime for managing the resources is seen as an efficient alternative to government management. A common property regime is defined as a situation in which individuals have mutual duties and rights. Individuals' right to limit the group size and right to prescribe the use of resource by each will together constitute the common property (Bromely, 1992). Experiences from many developing countries show that common property regimes for managing common pool resources like forests and fishing grounds are outcomes of collective action of local beneficiaries of the resource with the government playing a catalytic role. Some studies in India and other countries show that there are incentives to local people to manage local commons through a common property rights regime (Chopra et al., 1990; Murty, 1994). A classic example of common property rights regime is the management of fisheries (Dasgupta and Heal, 1979; Dasgupta, 1982). Fishing grounds as open access resources are normally over exploited with the number of fishing boats much above carrying capacity resulting in a commons type tragedy. This problem can be avoided by imposing catch quotas on fishermen on which they have to pay annual fee. The revenue from these quotas can be used to pay some fishermen for opting out from fishing. There can be incentives to fishermen to pay fees (receive compensation) so long as the net revenue (compensation) they receive is higher than the income they earn in an open access type of situation. This may result in a common property rights regime with the reduction of number of fishermen to the carrying capacity of the resource. These are some counterexamples to a proposition by Olson (1965) which states that voluntary collective action by individuals to achieve their group action is not possible and there is a need for coercion by an outside agency (government) to achieve the group interest.

In the developing countries where formal regulation of environmental externalities is weak or absent, instances of informal regulations in the form of Coasean bargaining between local communities and polluters are now found. Some recent studies have shown that there is growing evidence in Asia and Latin America about local communities taking recourse to social pressures, adverse publicity, threat of violence, recourse to civil law, and pressure through politicians, local administrators, or religious leaders for making polluters comply with the environmental standards. Pargal and Wheeler (1996) have shown that in Indonesia, the intensity of factory level water pollution has been highly responsive to characteristics of local communities, the factory level responsiveness increasing with the education levels, per capita income and employment status of local communities. Some recent studies (Misra and Murty, 1995; Murty and Prasad, 1995) of industrial water pollution abatement in India show that the establishment and success of common effluent treatment plants in industrial estates is related to the presence of a local NGO, possession by the local communities assets like farm land, milch animals, etc. which are affected by the water pollution and the education levels of local communities. One of the criticisms of the Coasean bargaining solution is that it is bilateral and in many real situations of environmental externalities, the bargaining can be between more than two agents. For example, the relevant agents for bargaining in the political economy of environmental pollution abatement are affected people, polluters, government and NGOs. However, in most of these cases the Coasean bargaining is still possible, if there is political organization of agents into two sub-coalitions and there is bargaining between sub-coalitions. For example, in the case of industrial water pollution abatement, there can be a coalition of affected people and NGOs and a coalition of factories and the bargaining between these coalitions can result in the efficient level of pollution. Therefore, in the case of externalities at a grander scale, it is bargaining with positive transaction costs that determines the efficient level of pollution which may not be Pareto optimal level as in the case of Coase's original proposition. Becker (1983) explains a Coase in politics situation by showing that there is no need of a benevolent government to control externalities like pollution, or even to collect taxes and make public spending; competition among politically-active pressure groups of people can produce the same result as a benevolent government.

Various institutional alternatives to government considered above for the control of environmental externalities contain some elements of market mechanism with the government playing a minimal to zero role. It looks paradoxical that in a case which is widely recognized as an example of market failure, one has to look now for market alternatives to deal with it. Given the doubtful quality of governments and high transaction costs of government instruments, it is imperative to look for new institutions to define and implement property rights for the environmental externalities so that market forces can deal with them efficiently. This phenomenon is reflected in the recent trend in the United States, Europe and other places in the increased role for private agents and agencies in the formulation and enforcement of environmental regulation. In the US, USEPA (US Environmental Protection Agency) increasingly relies upon private litigants to bring suits against polluters (Miller, 1988; Naysnerski and Tietenberg, 1992). In its 1993 Green Paper EC93, the European Commission clearly stated its intent to 'beep up' the rights of individuals and environmental groups to pursue for compensation and environmental damage done. Law Reforms Commission of Canada (1988) expressed its commitment to empowering private agents and agencies in almost all aspects of the pollution control process, from the drafting of regulations to the formulation of individual licenses and taking of enforcement actions. These developments can be seen as a move towards a free market solution to environmental problems.

Private action can be preemptive, if legal injunctions in civil courts are sought to prevent environmental damage occuring in the first place. In addition to civil courts, there are a number of other contexts such as licensing and judicial review of environmental regulation in which polluters and environmental groups 'contest' regulatory decisions. Both the agents can make efforts to lobby or capture regulators and politicians directly, or to apply popular pressure through media lobbying and direct action (Boehmer Christiansen, 1990). Such an activity may directly make regulations more stringent. The government may still have a role to play in an institutional setup in which polluters and environmental groups contest environmental regulations. By manipulating tax/subsidy to environmental groups, it may influence the outcome of the contest (Heys, 1997).

# 4. Collective Action for Industrial Water Pollution Abatement: An Institutional Alternative to Economic Instruments and a Benevolent Government[1]

SMITA MISRA and M. N. MURTY

There are many cases of externality management in developing countries which are carried out through contracts among the relevant agents for sharing the costs and benefits of controlling environmental externalities. Contracts can be found in the management of village commons and grazing lands, forest lands, fishing grounds, water resources management and even in the control of industrial pollution. Presence of incentives for all the relevant agents to cooperate in the management of an environmental externality is a precondition for such contracts to take place. For instance, incentives exist to have a contract between factories in an industrial estate and the affected neighbouring households for controlling water pollution if it results in reduction in damages to households and cost reductions in pollution control to factories.Collective action in environmental management is described as the emergence and execution of such a contract. This chapter provides various arguments in support of collective action in industrial water pollution abatement.

Section 4.1 describes collective action as an alternative institution to governmental instruments. Section 4.2 explores the possibility of having collective action in the developing countries and section 4.3 provides empirical evidence for the presence of scale economies in water pollution abatement which can make the collective action technology induced. Section 4.4 provides the conclusions.

[1] This chapter is drawn mainly from Misra and Murty, 1995.

## 4.1 COLLECTIVE ACTION AS AN INSTITUTION FOR ENVIRONMENTAL REGULATION

### 4.1.1 Conventional Instruments for Pollution Control

Environmental pollution caused by various economic activities operating in a free market is given as a classic example of market failure in welfare economics. Conventionally, governmental intervention in the market process in the form of pollution taxes, marketable pollution permits and CAC instruments is prescribed as a correction for market failure in controlling environmental pollution. The pollution taxes and marketable pollution permits can be designed in a manner such that the polluting economic activities can have incentives to optimally control environmental pollution by choosing cost minimizing technologies. In contrast, CAC instruments may not provide incentives to polluters to choose cost-minimizing pollution abatement technologies. Some empirical studies in the USA and other countries show that the pollution control costs under command and control instruments are several times higher than the least cost levels (Cropper and Oates, 1992). The optimal control of pollution through taxes or pollution permits requires both the pollution abatement cost function for generators and the damage functions of receivers of pollution to be convex. Given the convex abatement cost and damage functions, a tax on generators that is equivalent to the sum of marginal damages to receivers will provide incentives to generators for spending on pollution abatement, until the marginal cost of abatement is equal to marginal damages received by the affected parties (Baumol, 1972; Murty and Nayak, 1982). Even in the case of pollution taxes-standards or pollution permits-standards approach (Baumol and Oates, 1988), the abatement cost function should be convex to induce a generator to spend something on pollution abatement, so that, the prescribed environmental standards are realized through a tax or marketable permits.

The existing technology for pollution abatement, especially that of water pollution suggests that there are increasing returns to scale with respect to pollution loads, so that the abatement cost functions are non-convex. In other words, there is decreasing marginal cost in water pollution abatement, the presence of which will make pollution tax or marketable permits ineffective. Some empirical studies about the cost of the technological processes used for water pollution abatement confirm the presence of increasing returns to scale (Dasgupta and Murty 1985; Batstone et al. 1989; Mehta, Mundle and Sankar, 1993; Murty and James, 1996). The average cost of treatment falls as the quantum of wastewater treated increases, especially for volumes lower than the median sized volume in the industry.[2]

In the case of non-convex pollution abatement technologies, the pollution tax will not provide any incentives to the polluting firm to adopt pollution abatement technologies or spend something on pollution abatement. Since as shown in Fig. 4.1, the cost of pollution abatement (OFAE*) exceeds the total tax liability (OGBE*), the firm gains by paying the tax instead of spending on pollution abatement. Also, the government will not have an incentive to collect such a tax as it may not be in a position to recover the cost of pollution treatment with a Common Effluent Treatment Plant (CETP). Therefore, the incentives based on governmental instruments like taxes and marketable pollution permits fail to control water pollution. The only alternative to government in this case is to take recourse to CAC instruments which may compel the polluters to adopt some pollution abatement technologies. However, as pointed out earlier, the technologies used with CAC instruments need not necessarily be least-cost technologies.

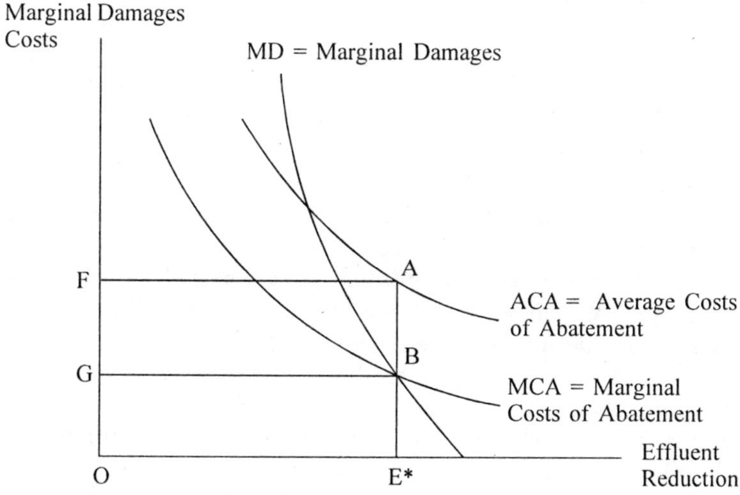


Fig. 4.1 Falling Marginal Cost of Abatement and Infeasibility of Pollution Tax

[2] Chapter 3 presents an estimation of a water pollution abatement cost function using data for an all-India sample of 131 factories drawn from 17 highly

Arguments for using economic instruments to control environmental pollution are mainly based on the assumption that government is benevolent. However, this may not always be the case. The government may be non-benevolent or the policing costs involved in dealing with externalities and public goods type of market failures. In such cases the only option left to the affected parties is to resort to collective action (Coase, 1960; Becker, 1983). Political influence exerted by the pressure groups of people can have similar effects as having a benevolent government to deal with market failures (Becker, 1983). Obviously there are transaction costs of collective action, but it can be shown that the welfare gains obtained through collective action are significantly higher than these costs (Murty, 1995b).

### 4.1.2 Collective Action of Polluters and Affected People

In cases where economic instruments are not effective in environmental pollution abatement, it is useful to look for institutional arrangements that facilitate collective action. For instance, Murty (1994a) has shown that collective action can result in the efficient management of common property resources. Also, it can efficiently control the externality problem in situations of non-benevolent government (Murty, 1995b). Collective action is possible if there are institutional arrangements in which all the relevant agents like polluters, affected people and the government, play an important role. We have the famous Coase theorem (Coase, 1960; Stigler, 1989) which can be stated in the context of water pollution abatement as follows: if there are legal institutions guaranteeing either the property rights of polluters to use water as a waste receptor or the property rights to affected parties to clean water, the costless bargaining by polluters and affected parties in a free market situation (zero transaction costs) can result in the first-best or Pareto efficient water quality standards and the final outcome of bargaining is invariant to the initial property rights assignment.

In Fig. 4.1 above, we can see that it is beneficial to pollutees to pay more than MCA to the polluter for every successive unit of improvement of environmental quality upto E*. Similarly, it is gainful to the polluter to pay the pollutees less than MCA for each successive unit of pollution until the environmental standard is

---

water polluting industries. It finds decreasing marginal cost with respect to volume of wastewater for given levels of concentrations of COD and BOD.

relaxed upto E*. The equilibrium level of environmental standard is E* which can be achieved through a bargaining process. Thus we have obtained an important result which shows that the Coase theorm holds good even with increasing returns to scale in the pollution abatement, while the pollution tax (Pigouvian tax) is ineffective in such a situation.[3] The assumption of zero transaction costs is crucial to the Coasean solution. This assumption implies that the damages due to pollution caused by an industry can be easily measured, compensation costlessly negotiated with all the affected people, and the payments can be fully enforced. In the case of a large number of affected parties, which is usually the case with environmental pollution, it is very difficult to sustain this assumption. There are significant transaction costs in the real world, and any institutional arrangement to deal with the environmental pollution abatement has to take cognizance of them. As pointed out earlier, the design and levying of a pollution tax which is the first-best instrument, has transaction costs. The original Coasean solution which assumes zero transacting cost is therefore not comparable to the pollution tax instrument. A modified Coasean framework which takes into account the transaction costs is comparable. The merit of the Coasean solution lies in dealing with two important situations: (a) the case of non-convex abatement cost function; and (b) the presence of a malevolent government.

An expansion of the Coasean theory to include collective action may make it possible to deal with the problem of environmental pollution abatement with positive transaction costs. The Coasean solution may be replicated through political competition among interest groups. Becker (1983) has suggested this 'Coase-in-politics' solution. He has suggested that the competition among political interest groups pursuing their self-interest provides government policies that help correct market failures. The high transaction costs of bargaining and paying compensation in the case of a large number of affected parties can be reduced to manageable levels, if the affected parties and polluters can organize themselves into clubs or interest groups.

[3] Note that the falling marginal abatement cost curve should intersect marginal damages curve from below for incentives to exist to either party for obtaining a bargaining solution. However, in a case in which falling marginal abatement cost curve intersects marginal damage curve from above, both economic instruments and Cosean bargaining fail to control water pollution optimally.

In the original Coasean framework, making a decision about the level of pollution and implementing it, is implicit in the contract arising out of bargaining. In actual situations, for example in the case of water pollution abatement using a taxes-standards method, the processes leading to fixation of pollution levels may be different from the processes aimed at implementing them. There can be collective action in the fixation of standards but implementation may be through economic instruments. In view of the ineffectiveness of economic instruments due to reasons cited earlier, there is a case for using Coasean bargaining approach both for fixation and implementation of standards. The case of water pollution abatement in small scale industries provides an example of this approach. In the case of many polluting firms with increasing returns to scale in pollution abatement, incentives can exist for firms to form a club for having a CETP with a cost-sharing arrangement.

In Fig. 4.2, $MCA_1$ and $MCA_2$ represent marginal abatement cost curves respectively in the cases of treating effluent by each polluter and by a CETP of a club of factories. If each polluter has to treat the effluent, the improvement in water quality is $E_1$ while in the case of CETP, the water quality to $E_2$ which is higher than $E_1$. In the former case, the total cost of improved water quality is $A_1B_1E_00$ with the shares of polluters and affected parties as $A_1B_1E_10$ and $B_1E_1E_0$. In the latter case, the total cost is $A_2B_2E_00$ which is lower than $A_1B_1E_00$. Also, in this case, both polluters and affected parties incur lower costs equivalent to $A_2B_2E_20$ and $B_2E_2E_0$.

A club of affected parties constituting the local communities can contribute to the control of pollution either by threatening the erring polluters with legal action or by setting up a treatment plant for the combined industrial and household effluents, and collecting the cost from the polluters. This arrangement is similar to the case of a municipal committee, consisting of representatives of local communities setting up a joint treatment plant for industrial and household effluents.Forming clubs of affected parties for water pollution abatement are feasible, if the effects of pollution are not dispersed geographically, and are confined only to local communities and settlements. In that case, the transaction costs or the costs of management of a club are less. Take, for example, a cluster of small industries polluting a lake or a river which is the source of drinking water for local communities. A club of local communities can set up a water treatment plant for drinking water and charge the cost to industries. Alternatively, the club can have an effluent treatment plant so that the water body is free from pollution. In this case, the water body apart from providing clean water for household and industrial use, protects various types of aquatic life in the lake and provides recreational facilities.

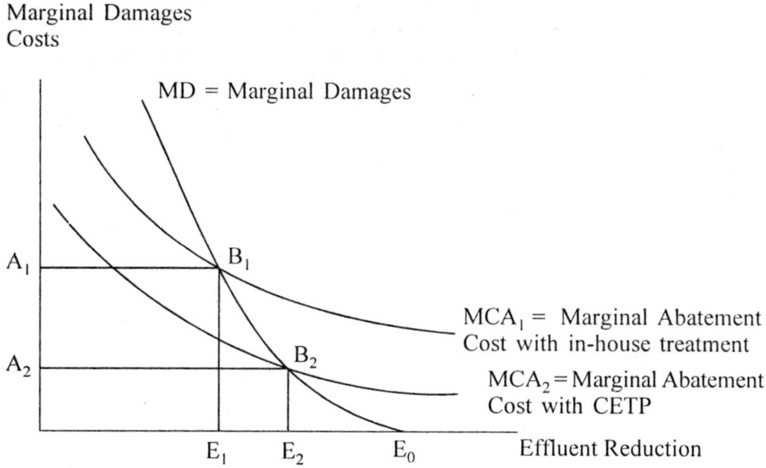


Fig. 4.2 Effect of Collective Action of Factories on Effluent Reduction

A study of developments leading to clubs or associations of factories and municipal committees setting up CETPs may provide a basis for the empirical verification of Coase-in-politics solution. Some studies have already provided empirical support to the extended Coasean theory of collective action. For example, Hamilton (1993) has shown that the installation of effluent treatment facilities in the USA is influenced by the presence of politically-active groups in a locality. Bargaining between the clubs of factories and the municipal committee consisting of representatives of local communities may result in the setting up of a CETP by either of the bargaining parties. However, unlike the original Coasean solution, it is a case of bargaining with positive transaction costs. In this case, there are transaction costs in the form of the cost of forming a club or a committee and the threat of legal action in the case of non-compliance, by either party of the agreement arrived at through bargaining. It is now known that the Coase theorem, in its original form, no longer holds with positive transaction costs. If one recognizes the income effects and positive transaction costs of bargaining, the outcome of bargaining, that is, the environmental quality, depends upon the initial property rights assignments. Ignoring the income effects, let us consider only the effect of positive transaction costs on the final outcome of bargaining. If the factories have the property right to pollute water resources, and the affected parties do not have the right to clean water (zero liability rule), the marginal net benefit to the affected parties or damages avoided from the pollution abatement will be less with positive transaction costs. As shown in Fig. 4.3 and 4.4 below, the environmental quality in this case is $E_1$ which is lower than $E^*$, the environmental quality with zero transaction costs.

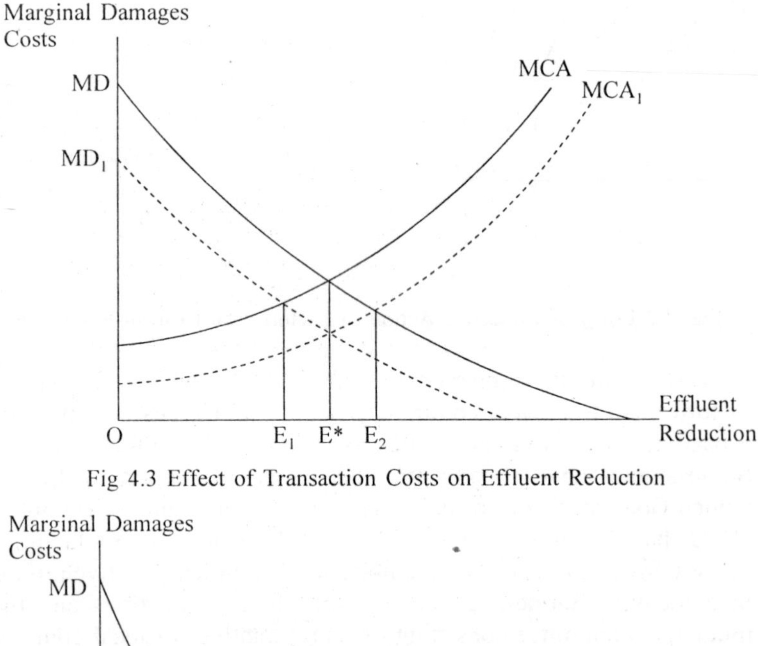


Fig 4.3 Effect of Transaction Costs on Effluent Reduction

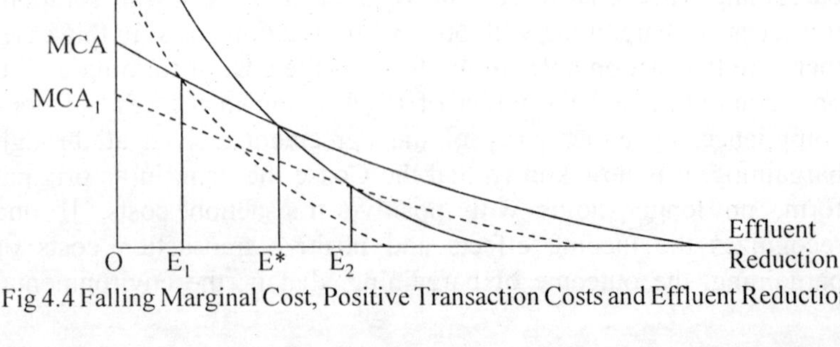


Fig 4.4 Falling Marginal Cost, Positive Transaction Costs and Effluent Reduction

Note that in these figures,

| | | |
|---|---|---|
| MD | = | Marginal damage with zero transaction cost; |
| $MD_1$ | = | Marginal damage with positive transaction cost (zero liability law); |
| MCA | = | Marginal cost of abatement with zero transaction cost; and |
| $MCA_1$ | = | Marginal cost of abatement with positive transaction cost (full liability law) |

Alternatively, if the affected parties have the right to clean water (full liability rule), the marginal cost of abatement will be less with positive transaction cost. In this case, the environmental quality is $E_2$ which is higher than $E^*$. These results hold good in both cases of convex and non-convex abatement technologies as shown in Figs 4.3 and 4.4. In the case of usual convex technologies of pollution abatement, the choice between Coase-in-politics solution and the pollution (Pigouvian) tax is purely guided by the transaction costs in each case. Choice between these two alternative institutional arrangements has to be made by comparing the costs of collecting information about abatement cost and damage functions and monitoring the compliance with the tax instrument by industries, with the costs of organization of political groups of clubs of polluters and clubs of affectd parties/municipal committees and the legal action that may be resorted to by each group. However, in the case of non-convex abatement technologies, there is no alternative to collective action. Even in the case of convex technologies, if the government is malevolent or the regulator is corrupt, collective action is the only institutional alternative to efficiently control externalities arising out of water pollution.[4]

## 4.2 COLLECTIVE ACTION IN DEVELOPING COUNTRIES: IS IT POSSIBLE?

The Coasean bargaining approach suffers from a number of limitations restricting its practical relevance, especially for a developing economy. These limitations pertain to cost incurred in reaching the

[4] In a recent paper, Murty (1995b) has shown that if the government is non-benevolent thus making economic instruments ineffective, collective action or political organization of affected parties can optimally control industrial water pollution.

Coasean solution, bargaining power of groups involved and equity considerations. As explained earlier, Coase assumes costless market transactions. However, in the real world and especially in the developing economy context, the cost of altering and combining rights through the market to achieve efficient bargaining might be so great, that the optimal arrangements of rights and the consequent preservation of the environment may never be reached. The State then plays an important role. It designs the initial allocation of rights, and by doing so, might incur lower transaction costs than private agreements.

The Coase theorem also assumes complete information. When there are information asymmetries, the parties will have incentives to cheat and reveal false information, thereby making bargaining inefficient. Thus, it is necessary for the government to induce people to reveal true information through an incentive scheme, and to render participation to the scheme compulsory (Farrell 1987). Farrell argues that such a scheme might not be efficient; but then the comparison is between the inefficient outcome of a private bargaining process and the inefficient outcome of an uninformed central authority. He shows that while 'the bumbling bureaucrat outperforms both allocation of rights', an inefficient centralized decision-making process yields a better outcome than private bargaining.

Another important problem is that, it is not clear what role bargaining power plays in the Coasean framework. All parties concerned in the Coasean framework should be in a position to negotiate. However in the water pollution abatement case, it is the future generation who bear the cost of the externality, and it is not clear how their interests may be accounted for in the bargaining process. Also, Coase's Theorem bears only on the issue of efficiency. The distributive effects of private bargaining are not considered.

While the Coasean solution (which might even be a Nash equilibrium) may be at one end of the spectrum with self-enforcing behaviour of individuals, contractual arrangements with complete state regulations could be at the other end. Self-interest drives the former while an external enforcer drives the latter. The latter framework depends on the extent of market distortions and the failure of Coasean assumptions.In between would lie a whole range of arrangements to influence the outcome, some of which are highlighted here in the context of water pollution abatement in developing countries.

Collective action in industrial water pollution abatement is meant for bringing necessary institutional changes that are compatible with the choice of cost-saving technologies in comparison to the technologies chosen in the free market. In the case of market failures, as is the case with water pollution and a situation of non-benevolent government, these institututional changes have an instrumental value for facilitating the choice of efficient technologies. Empirical evidence about water pollution abatement in India shows that collective action takes place if the cost-saving technological options in relation to the ones chosen in the free market are available and changes in the social institutions that facilitate the choice of such technologies are possible. Given better technological options than the ones chosen in the free market, collective action may be undertaken to bring about the necessary institutional changes to facilitate their choice. Foster (1981) has rightly observed that 'the social problems can be solved only by adjusting institutional structures involved in the problem so as to bring them into instrumentally-efficient correlation with the technological aspects of the problem'. For example, the technology CETP for an industrial estate can be adopted if necessary legislations are made to define the property rights of factories and affected people and the active political groups of clubs of factories and the associations of affected parties are formed. The emergence of active political groups depends upon the benefits each group gets from the bargaining. A CETP for an industrial estate confers benefits of saving in costs to factories and reduction in damages to the affected parties. Also, the extent to which parties concerned about industrial water pollution act collectively depends upon the degree to which the activities of one party affect the other, and vice versa. More education and participation of people and industrial concerns are needed, if the institutional adjustment is to fulfill its instrumental potential to choose efficient technologies.

Collective action in industrial pollution abatement may not be possible due to various factors that constrain the relevant agent's willingness and ability to participate. Some of these constraints are

- Inadequate and ambiguous environmental law protecting the rights of affected parties to a clean environment.
- Lack of public awareness about the magnitude of damages to the community from pollution and ignorance about the existence of environmental law for protecting the rights of people.

• Lack of resources of local communities to organize themselves as politically-active groups for taking recourse to legal action against polluters.

• Inadequate resources at the disposal of small enterprises to organize themselves as a club for setting up a CETP.

• The problem of agreeing upon a cost sharing arrangement for a CETP by a club of factories.

• The problem of designing methods of charging factories and households for treating effluents by a CETP set up by a club of affected parties/municipal committee.

• Political uncertainty in the case of clubs of affected parties and the uncertainty of continuation of the agreement for cost sharing by a club of factories.

The government can play an enabling or catalytic role in the collective action that aim to remove some or all of these constraints. In a democracy, the legislatures can enact environmental laws and create legal institutions to protect the rights of citizens to a clean environment. Public awareness about the magnitude of damages from pollution can be promoted by creating institutions for environmental education. The government or NGOs can even provide a subsidized legal facility to local communities. It can give small enterprises concessional financial loans and technical help for setting up CETPs. NGOs can provide expertise to a club of factories to work out mutually agreeable methods of sharing the cost of CETP.[5]

There are many incentives for polluters, affected parties and the government to promote collective action in industrial water pollution abatement:

*Incentives to polluters*

1. CETP is less expensive in terms of capital and operation cost.

2. It is easy for a club to secure financial support from the government and NGOs.

3. With a CETP, water can be treated economically to produce process-grade water which can be reused and recycled in the

[5] A method of sharing the cost of CETP is an important component of contract arising out of collective action of small scale enterprises in an industrial estate. Therefore, the method of cost sharing should be such that it is mutually acceptable to all enterprises. Appendix A7 presents some such cost sharing methods.

industry. It is an important incentive to form a club in water-scarce regions.

4. The size of an industrial estate depends upon water availability and the facility to dispose wastewater on a sustainable basis. CETP may help achieve both these objectives.

*Incentives to affected parties*

1. Improved quality of drinking water
2. Reduction in damages from water-borne diseases
3. Recreational facilities from the preserved water body
4. Reduction in the cost of legal action against polluters
5. Increase in access to legal institutions

*Incentives to government*

1. Reduction in the burden of various government agencies working for abating and controlling water pollution.

2. Respectability earned by government agencies for the catalytic role, in contrast to unpopularity for the coercive role.

The observed complementarity between the collective action in industrial water pollution abatement and the interests of all the relevant parties indicates that the collective action involving factories, affected parties and the government is feasible. It also highlights the roles of different parties, especially the catalytic role of government.[6]

## 4.3 SCALE ECONOMIES IN WATER POLLUTION ABATEMENT: A CASE STUDY OF INDUSTRIES IN HARYANA

### 4.3.1 Current Environmental Policy and Small Enterprises

More than twenty years of environmental legislation and institutional development in India has resulted in the use of CAC instruments for pollution control in both big and small factories. The Government of India has been providing certain fiscal incentives in the form of direct and indirect tax concessions on the pollution control equipment,

[6] Murty (1994a) has shown that a similar type of complementary relationship exists between collective action for the preservation of local commons and forest lands and the interests of local communities and the government.

the benefits from which accrue mainly to the big factories. Since the accrual of benefits from fiscal incentives *ipso facto* depends upon the installation of pollution control equipment by the factory, small factories are not able to reap these benefits due to inability to mobilize resources for setting up their own effluent treatment plant. Also fiscal incentives or subsidies to the polluting factories (especially the big factories) are not desirable from the point of view of allocative efficiency since they encourage the factories to increase their scale of operation and hence the level of pollution (Baumol and Oates, 1988; Murty,1994b). Ideally, it is the pollution tax on big factories that induces them to use the least-cost pollution abatement technologies. But the Government of India so far does not have an instrument like a pollution tax. The currently levied water cess on the industrial use of water is very nominal and can, at most, be regarded as a revenue tax.

Use of CAC instruments by the government in an environment of non-availability of economically viable technological options for pollution abatement has caused considerable hardships to small-scale enterprises. Given that the government-managed public sector has been the fountainhead of industrial development in India during all these years, the government has not made any sincere efforts to promote economically viable pollution abatement technologies for the small-scale enterprises via investments in R&D in the public sector. Some efforts have been made only recently, especially with the financial support of international agencies like the World Bank, to promote pollution abatement technologies like the CETP for the small scale enterprises.

The threat of closure and penalties for the non-compliance with the prescribed pollution standards has resulted in the widespread use of less cost-effective or inefficient pollution abatement technologies by the small scale enterprises. The inability to make necessary investments in the pollution abatement has resulted in the closure of several small-scale enterprises in various industrial estates in India. The data about the investment in the effluent treatment plant (ETP) for some of the small-scale enterprises in the Sonepat industrial area show that the cost of ETP ranges between 8 to 10 per cent of the cost of the plant as shown in the Table 4.1. The escalation of the production cost due to involuntary use of economically non-viable pollution abatement technologies can make it difficult for the small enterprises to compete with their bigger counterparts. It can especially happen when there are scale economies in the pollution abatement and the government uses coercive measures like CAC instruments. This problem can be solved if the Government plays a catalytic role in the promotion of cost effective pollution abatement technologies like CETPs for the industrial estates containing small-scale enterprises and the creation of legal institutions for the promotion of collective action in pollution abatement.

Table 4.1
Fixed Costs of Main and Effluent Treatment Plants for Small-Scale Enterprises in the Sonepat Industrial Area (Rupees Lakhs)

| Industrial Unit | Fixed Costs of the Main Plant | Fixed Costs of the Effluent Treatment Plant | Effluent Treatment Costs as a percentage of Main Plant Costs |
|---|---|---|---|
| Bharat Leathers | 80.00 | 13.269 | 17 |
| Jai Shree Industries | 24.00 | 4.934 | 21 |
| Indian Leathers | 31.00 | 18.000 | 58 |
| Sugra Leathers | 300.00 | 39.236 | 13 |
| Kiran Overseas (incl. Rahul Leather) | 500.00 | 40.000 | 8 |
| HBR Chemicals (incl. Supreme Organics) | 25.25 | 27.000 | 107 |

*Source:* Survey Data.

### 4.3.2 Water Pollution Abatement Practices in the Sonepat Industrial Area

The Sonepat industrial area harbours a large number of factories belonging to industries of Dye Stuffs and Chemicals, Pharmaceutical, Tanneries, Textiles, Electroplating, Engineering, etc. With the exception of a few big factories, most of these factories are of small and medium size and hence are unable to deal with the problem of water pollution abatement due to various reasons. For studying the problem of industrial water pollution in this area, we have surveyed most of the industrial units belonging to Leather and Leather Processing, Dye Stuffs and Chemicals and Pharmaceutical. Tables 4.2 and 4.3 describe the production practices of the various industrial units surveyed.

Out of eight tanneries surveyed, four produce the finished leather, one produces only semi-finished leather, while the remaining three produce both semi-finished and finished leather. The installed

Table 4.2
Production and Cost Structure of Tanneries in the Sonepat Industrial Area

| S. No. | Commodity | Daily Production Capacity | Actual Daily Production | Percentage of Actual to Capacity Production | Fixed Costs (Rs Lakhs) | Employees (Number) |
|---|---|---|---|---|---|---|
| 1. | Finished Leather | 2,000 | 1,500 | 75 | 80 | 100 |
| 2. | Semi-Finished Leather | 600 | 300 | 50 | 8 | 15 |
| 3. | Semi-Finished & Finished Leather | 7,000 | 6,300 | 90 | 24 | 35 |
| 4. | Finished Leather | 260 | 260 | 100 | 6 | 15 |
| 5. | Semi-Finished & Finished Leather | 700 | 500 | 71 | 31 | 12 |
| 6. | Semi-Finished & Finished Leather | 1,000 | 500 | 50 | 20 | 5 |
| 7. | Finished Leather | 5,000 | 2,500 | 50 | 300 | 100 |
| 8.. | Finished Leather | 6,000 | 4,000 | 67 | 500 | 200 |

*Source:* Survey Data.

Table 4.3
Production and Cost Structure of Chemical Units in the Sonepat Industrial Area

| Industrial Unit | Capacity (mt/month) | | (Actual/Installed)*100 | Fixed Cost (Rs Lakhs) | Number of Employees |
|---|---|---|---|---|---|
| | Installed | Actual | | | |
| Ashoka Chemicals | 5 | 4 | 80 | 10 | 8 |
| Modern Organics | 5 | 3 | 60 | 10 | 17 |
| HBR Chemicals | 50 | 40 | 80 | 22 | 74 |
| Master Industries | 2 | 2 | 100 | 9.42 | 12 |
| Orient Industries | 5 | 3 | 60 | 5.8 | 7 |
| P.D Industries | 3 | 2 | 67 | 3 | 10 |
| Metachem Industries | 30 | 20 | 67 | 41 | 60 |
| Supreme | 3 | 2.5 | 83 | 3.25 | 32 |

*Source:* Survey Data.

capacity of tanneries varies from 260 pieces per day to 7,000 pieces per day. Their capacity utilization varies from 50 to 100 per cent. The availability of raw hides is an important hurdle for their full capacity operation. The fixed cost and employment of these units are respectively in the ranges of Rs 6,500 lakhs and 5–200 people. Out of eight chemical plants surveyed, two have an installed capacity of 30 metric tonnes per month while the others have a capacity of less than or equal to 5 metric tonnes per month. The capacity utilization varies from 60 to 100 per cent. Many of these units are very small, with the fixed cost varying from Rs 3 to 40 lakhs and employment of 7 to 74 persons.

The industrial units surveyed in the Sonepat industrial area have revealed varied pollution abatement practices to meet the Haryana pollution control standards. Almost all of the bigger units have a complete ETP which can help in meeting the prescribed effluent standards. Most of the smaller units have an evaporation tank which can help to meet only primary treatment requirements. Some of the smaller units have yet to construct the evaporation tanks.

Out of eight tanneries surveyed in Sonepat, five have a complete ETP. All these five units have high fixed costs. The remaining three units, that is Nanda Leathers, Bharati Transnational and Santosh Paragal & Co, are much smaller units with fixed costs of only Rs 6–20 lakhs (Table 4.2). At present none of these are meeting the required pollution control standards. The evaporation tanks for some of these units are under construction with an estimated expenditure of Rs 2 lakhs each. The capital cost of treatment as a proportion of the capital cost of the main plant generally turns out to be much larger for the smaller units (Table 4.1). Since they are situated close to each other, they are ideally suited for a CETP, enabling them to economize on their overhead costs and space.

A survey of nine chemical units has revealed that only two chemical units had a joint ETP to meet the Haryana Pollution Control Standards. The remaining seven units are functioning with an evaporation tank to meet the primary treatment requirements. The joint ETP of HBR and Supreme Chemicals has excess capacity and is ideally located to accommodate the effluents from PD Metachem and Master Chemicals. This could be an instance of having a big factory as a leader in promoting a CETP with lower per unit cost of treatment for all.

The survey shows that only those factories which have a complete ETP are able to meet the Haryana Pollution Control Standards of

30mg/l BOD, 250mg/l COD, 100mg/l SS and 9 PH. The others either already have or are proposing to build an evaporation tank to meet the primary treatment requirements. In the case of dye units also, only the HBR and Supreme Chemicals are able to meet the Haryana Pollution Control Standards with their joint ETP. The other units also must treat their effluents independently or collectively. Tables 4.4 and 4.5 provide details about the water consumption and residual water and its characteristics for the industrial units surveyed in Sonepat.

The estimates of annualized opportunity cost and financial costs of treatment per kilolitre of residual water incurred by the tanneries in the Sonepat Industrial Area have been worked out on the basis of the standard practices. The annual capital cost is estimated as 23 per cent of capital cost (i.e. 18 per cent rate of return on investment + 5 per cent rate of depreciation) plus 18 per cent of the cost of land and contingencies. The financial cost is estimated with the assumption of 14.5 per cent interest rate and depreciation of 5 per cent. The opportunity and financial cost estimates for the tanneries surveyed exhibit decreasing average costs. The average cost in general decreases as the volume of the effluent increases as shown in Table 4.6 and Fig. 4.1. The financial cost of treatment per kilolitre of residual water decreases from Rs 46 to Rs 19 as the volume of residual water increases from 803 kl to 54750 kl. This supports our basic proposal for having a CETP for small-scale units with decreasing costs in water pollution abatement technology.

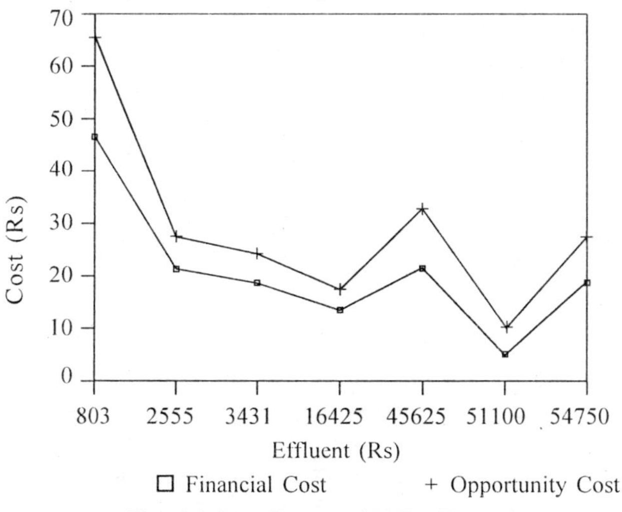


Fig. 4.5 Cost Curves of ETPs (Tanneries)

Table 4.4

Residual Water Characteristics of Tanneries in Sonepat

| Industrial Unit | Daily Water Intake (KLD) | Daily Residual Water (KLD) | Residual Water Characteristics | | | | | | | |
|---|---|---|---|---|---|---|---|---|---|---|
| | | | *Before Treatment* | | | | *After Treatment* | | | |
| | | | BOD | COD | SS | PH | BOD | COD | SS | PH |
| Bharat Leathers | 150 | 140 | 110 | 368 | 124 | 7.2 | 28 | 136 | 61 | 7.4 |
| Nanda Leathers | 10.5 | 9.4 | 280 | 640 | 970 | 4.1 | | | | |
| Jai Shree Industries | 51.0 | 45.0 | 105 | 448 | 864 | 11.8 | 26 | 148 | 83 | 8.3 |
| Bharati Trans. | 2.2 | 2.2 | 375 | 978 | 328 | 3.5 | | | | |
| Indian Leathers | 23.0 | 22.0 | 360 | 864 | 683 | 9.9 | 28 | 112 | 63 | 7.5 |
| Santosh Pargal | 7.0 | 7.0 | 395 | 1248 | 872 | 7.62 | | | | |
| Sugra Leathers | 154.0 | 125.0 | 180 | 570 | 210 | 5.2 | 27 | 115 | 70 | 6.8 |
| Kiran Overseas | 160.0 | 140.0 | 88 | 284 | 290 | 7.4 | 28 | 112 | 90 | 7.4 |

*Source:* Survey Data.

*Note:* KLD: Kilolitres per day.

Table 4.5

Residual Water Characteristics of Dyestuff Units in Sonepat

| Industrial Unit | Daily Water Intake (KLD) | Daily Residual Water (KLD) | Residual Water Characteristics | | | | | | | |
|---|---|---|---|---|---|---|---|---|---|---|
| | | | *Before Treatment* | | | | *After Treatment* | | | |
| | | | BOD | COD | SS | PH | BOD | COD | SS | PH |
| Ashoka Chemicals | 11.5 | 10.20 | 140 | 512 | 142 | 7.2 | | | | |
| Modern Organics | 0.85 | 0.60 | 180 | 976 | 1190 | 10.5 | | | | |
| HBR Chemicals | 30.00 | 23.00 | 115 | 413 | 564 | 7.4 | 24 | 158 | 78 | 6.9 |
| Master Industries | 1.40 | 1.30 | 400 | 600 | 150 | 4.0 | | | | |
| Master Chemicals | 3.50 | 2.40 | 140 | 416 | 6296 | 7.2 | | | | |
| Orient Industries | 5.80 | 3.80 | 360 | 9960 | 80 | 6.6 | | | | |
| P.D Industries | 2.40 | 2.00 | 330 | 2880 | 142 | 2.0 | | | | |
| Metachem Industries | 17.00 | 12.80 | 500 | 800 | 220 | 5.0 | | | | |
| Supreme | 2.20 | 1.80 | CETP with HBR Chemicals | | | | | | | |

*Source:* Survey Data.

The survey of some industrial units in Sonepat exhibits a clear possibility of collective action as an effective instrument for controlling water pollution in the industrial estates. Individual producers are very keen to adopt the CETP technology. Nanda Leathers, Bharati Transnational and Santosh Pargal have individually given their consent for a CETP. However they are attributing competition and lack of cooperation as the two main factors hindering the CETP technology. The survey of chemical dye plants clearly brings out the case of a big factory as a leader in promoting the CETP. The joint ETP of HBR and Supreme Chemicals as a leader in promoting a CETP can easily accommodate the effluents from P. D., Metachem and Master Chemicals. This could be an instance of a big factory acting as a leader in promoting a CETP with a lower per unit cost of treatment for all.

Table 4.6
Annualized Financial Costs and Opportunity Costs of Water Pollution Abatement for Tanneries (1993–4 Prices)

| | Residual Water (kl) (1) | Opportunity Cost (Rupees lakhs) (2) | Financial Cost (Rupees lakhs) (3) | (2)/(1) (Rupees) (4) | (3)/(1) (Rupees) (5) |
|---|---|---|---|---|---|
| 1. Bharat Leathers | 51100 | 4.002 | 2.874 | 8 | 6 |
| 2. Nanda Leathers | 3431 | 0.801 | 0.631 | 23 | 18 |
| 3 Jai Shree Industries | 16425 | 2.635 | 2.215 | 16 | 13 |
| 4. Bharat Trans. | 803 | 0.540 | 0.370 | 67 | 46 |
| 5. Santosh Pargal | 2555 | 0.714 | 0.544 | 28 | 21 |
| 6. Sugra Leathers | 45625 | 13.560 | 10.225 | 30 | 22 |
| 7. Kiran Overseas | 54750 | 13.580 | 10.180 | 25 | 19 |

*Source:* Survey Data.
*Note:* Column (4) shows the opportunity cost per kl of residual water; Column (5) shows the financial cost per kl of residual water.

The Government will have to play a catalytic role in removing the bottlenecks and in promoting the CETP technology.

The Sonepat survey reveals the following advantages of a CETP technology:

1. The small tanneries and dyestuff units operate on the basis of 'one shift a day' generally employing less than 20 labourers. Hence the maintenance of an ETP would be difficult for them. All the units

surveyed certainly prefer a CETP whereby their operating costs would be lowered.

2. The fixed cost of the ETP is very high (as compared to the fixed cost of the plant). Thus the CETP would ensure savings in the capital cost for small-scale units which lack finance. A single CETP would be cheaper than a multitude of scattered smaller unit ETPs.

3. The smaller tanneries and dyestuff units surveyed suffer from a shortage of space. They would gladly adopt the CETP technology and transport effluents to the CETP.

4. The technologies adopted by the small-scale units are mostly outdated. Simply by constructing an evaporation tank they cannot meet the required pollution control standards. Specialisation in CETP technology would ensure a better quality effluent.

5. The regulation and monitoring work of pollution control would be considerably reduced if CETPs are implemented. The CETP would ensure an organised and streamlined effort.

The above reasons support the case for a CETP. It would not only be cost-effective, but would also help the regulatory authority in enforcing the standards.

## 4.4 CONCLUSIONS

There are limitations on conventional instruments like pollution taxes, marketable permits and CAC instruments for controlling industrial pollution, especially if there are non-convexities in pollution abatement and government is non-benevolent.

Collective action *a la* Coase and Becker can optimally control water pollution even with convex pollution abatement technologies. Collective action is possible if there are institutional arrangements in which all the relevant agents, namely, polluters, affected parties and the government can play a role. The bargaining solution of Ronald Coase can be generalized to include collective action which can make it possible to deal with the problem of environmental pollution abatement with positive transaction costs. The high transaction costs of bargaining in the case of a large number of polluters and affected parties can be reduced to manageable levels if the polluters and the affected parties can organize themselves as clubs or politically-active groups.

Collective action is an alternative institution to the government for environmental regulation. The role of collective action becomes important when the Government is non-benevolent. Also, it results in savings in cost of regulation because it facilitates the decentralization of environmental management.

Collective action requires the government to play a minimal role of creating the initial property rights for the environmental resources. Collective action of affected people and the bargaining between a coalition of polluters and a grand coalition of affected parties can result in the optimal control of environmental pollution.

Empirical evidence about the presence of scale economies in water pollution abatement by industries suggests that it is economical for small-scale industries to control water pollution through collective action. Small-scale enterprises cannot reap the benefits of the fiscal incentives provided by the government for the control of pollution since *ipso facto* the accrual of benefits depends upon the installation of pollution control equipment by the factory and small-scale enterprises cannot mobilize the resources to have their effluent treatment plants. Use of command-and-control instruments have compelled the small-scale enterprises to use economically unviable pollution abatement technologies and to face the threat of closure in case of non compliance of prescribed pollution standards.

The Sonepat survey reveals that small-scale industries like tanneries and chemical dye units have scale economies with respect to water pollution abatement. The cost of treatment per kilolitre of residual water decreases as the volume of residual water increases, justifying the case for having a CETP for small-scale enterprises.The larger tanneries and chemical plants with high initial fixed costs are meeting the pollution control standards with ETPs of their own. The CETP technology would enable the small-scale industries also to meet the pollution control standards.

Individual producers support collective action via CETP in water pollution abatement. They feel that it would not only be cost-effective, but will also ensure a better quality effluent with superior technology. The CETP technology will make the job of the regulatory authority in enforcing the standards easier.

Nearly one-fifth of the World Bank loan for promoting industrial pollution control practices in India is earmarked for the development of CETPs in industrial estates. This would be supported by matching grants from the Government of India, state governments, domestic financial institutions and associations of factories.

# 5. Emissions Reduction and Influence of Local Communities in India

M. N. MURTY and U. R. PRASAD[1]

There is now growing evidence in the developed as well as developing countries about the influence of local communities on industries in reducing pollution.[2] There is a need for local communities to influence the pollution control decisions of industries even with the presence or absence of formal environmental regulations. There is now considerable evidence about instances in which the fixation of local environmental standards and compliance by industries are achieved through Coasean bargaining between industries and local communities. When formal regulation is weak or absent, informal regulation through local community pressure can make the industries undertake pollution abatement. Even if formal regulation is present, the factories and environmental groups can contest the regulatory decisions of government through lobbying and political influence (Heyes, 1997).

The amount of influence local communities can exert on the factories to undertake pollution abatement depends, among other factors, upon their affluence, the degree of political organization, education and environmental awareness.[3] Some recent studies about the role of local communities in controlling environmental externalities

[1] This chapter is drawn mainly from Murty and Prasad (1997).

[2] For the East Asian and Japanese experience see O'Connor (1996) and Murty (1995a)

[3] For example Pargal and Wheeler (1996) observe that a community's 'environmental supply schedule' depends on characteristics such as income, education, level of civic activity, legal or political recource, media coverage, presence of a non-governmental organization, the efficiency of existing formal regulation, and the total pollution load faced by the community in relation to the absorptive capacity of the local environment.

have produced mixed results. For example, a study by Pargal and Wheeler (1996) has found a negative relationship between BOD load in a factory effluent and per capita income and education level of local communities in a sample of 243 factories in Indonesia, while the study of Pargal, Mani and Huq (1977) has found no significant relationship between BOD load and development index[4] of local districts in a sample of 250 factories belonging to leather, oil and petro-chemicals, paper, iron and steel, synthetic fibre and textiles in India. Some recent studies in water pollution abatement in industrial estates in India (Murty and Prasad, 1995; Misra and Murty, 1995b) provide evidence of local communities playing a significant role in pollution abatement.

In developed countries, contesting government environmental regulations by local communities or environmental groups and industry is now becoming an important component of enforcement of environmental laws. Some studies (Naysnerski and Tietenberg, 1992; Miller, 1988) have reported that the US Environmental Protection Agency (USEPA) encourages private litigants to bring suits against polluters. Private litigation now constitutes a major component in the enforcement of US Clean Air Act. The European Commission in its 1993 Green Paper, has clearly signalled its intent to beef up the rights of individuals and environmental groups to pursue polluters for compensation for environmental damage done (Heyes, 1997). Baik and Shogren (1994) have reported that the law-makers in Canada are committed to empowering private agents and agencies in virtually every aspect of the pollution-control process, from the drafting of regulations to the formulation of individual licences and undertaking enforcement actions (Law Reforms Commission of Canada, 1988). These developments are now popularly seen as moves towards free market solutions to growing environmental problems. A factory and a private agent (an individual, some coalition of individuals or an environmental group) contest a regulatory decision. This may, for example, be the decision regard-

[4] Relative development index: Given the non-availability of comparable district income data, CMIE has computed an index on the relative development of a district in India as a proxy indicator. The weighting pattern adopted to develop this index reflects, by and large, the importance of the different sectors of the economy. The agriculture sector, the mining and manufacturing sector and service sector are respectively given the weights 35 per cent, 25 per cent and 40 per cent.

ing whether or not to oblige a firm to locate itself at a particular site or to force a firm to clean up a damaged site. The benefit to the private agent is a part or whole of the monetary value of the environmental damage. The firms contest environmental decisions of governments in courts and make efforts to win cases against those decisions which involve very high compliance costs.[5]

India, unlike many other developing countries, has a comprehensive set of environmental laws as explained in Murty (1995a). However, various agencies implementing these laws are only partially successful in bringing about the required level of preservation of environmental resources.[6] There are both success and failure stories of monitoring and implementation of environmental laws in India. For example, in the case of industrial pollution abatement, there are certain industrial locations in which environmental regulations are successful, while in many other locations they are completely ineffective. Therefore, the role of local communities in industrial pollution abatement becomes important in both the contexts described above: (a) environmental regulations are absent or completely ineffective and there is Coasean bargaining between the factory and local community and (b) the environmental regulations are present and are contested by the local communities and factories.

## 5.1 MONITORING AND ENFORCEMENT OF ENVIRONMENTAL REGULATIONS

In India, there is scope for local communities to influence decisions affecting environmental quality in the spheres of licensing and locational choices of developmental projects, fixation of local environmental quality, monitoring and enforcement of environmental regulations, and environmental impact assessment. The government plays a catalytic role by sponsoring programmes for promoting the participation of local communities and environmental groups in environmental management. Local communities can play a pivotal role by contesting in civil courts the activities of the government in monitoring and enforcement of environmental regulations and industry in complying with environmental standards.

[5] See Heyes (1997) for details about how effective environmental decisions are determined as the outcomes of 'contests' between private agents and firms.

[6] See also Murty (1997) for the details about the current state of implementation of environmental laws in India.

Monitoring and enforcement of environmental regulations in India has been partially successful so far. Enforcement provisions of environmental laws in India provide for severe punishments leading upto seven years of imprisonment for the non-complying party. The number of court cases launched under the Water and Air Acts of 1974 and 1977 by the State and Central Pollution Control Boards has been steadily rising. They have increased from 2000 in 1987 to over 6000 in 1996. There is considerable delay in clearing these court cases, with the number of cases pending in courts rising continuously. The court decisions in most of the cases are in favour of pollution control boards, with 60 to 80 per cent of cases going in favour of the boards. During a recent year, conviction orders under the Water Act amounted to 109 imprisonment orders, 72 orders for fines and 173 restraint orders. In the case of the Air Act out of 293 convictions in 1992, 153 were imprisonment orders, 138 decisions were fine orders and only 2 were restraint orders.

The effectiveness of enforcement and monitoring of environmental regulation can be judged by examining the ambient air and water quality, the status of forest resources, wildlife, biodiversity etc, in different parts of the country during the period in which environmental regulation is in operation in India. Atmospheric pollution, as indicated by various parameters of air pollution in the four major cities of Delhi, Bombay, Calcutta and Madras shows a mixed trend during the period 1978–90. Atmospheric pollution has increased with respect to all four parameters of SPM (suspended particulate matter), $SO_2$ (sulphur dioxide), $NO_2$ (nitric oxide) and P-index[7] in Delhi, with respect to two parameters SPM, $NO_2$ in Bombay and with respect to one parameter $NO_2$ in Calcutta. However, in Madras there are significant reductions in the atmospheric pollution with respect to all four parameters. In the case of monitoring of ambient water quality, the number of monitoring points in the country have increased from 9 in 1979 to 324 in 1989. During this period, the percentage of number of monitoring points at which BOD standard is violated, declined from 66.7 to 46.6 per cent, and that at which the coliform standard in terms of the permissible amount of faecal coliform bacteria in a unit of water is violated, declined from 44.4 to 43 per cent. There was a marked reduction in the BOD level of the river Ganges during the period 1980–9 with the degree of

[7] P-index is an air quality index or pollution index for air quality.

violation of standards falling from 7.80 to 2.21 (which can be attributable to a major project called Ganga Action Plan started in 1985) while in the other rivers, the degree of violation of BOD standard has fallen from 4.21 to 2.83 during the same period. The Annual Report of the Central Pollution Control Board (CPCB) for the year 1994–5 (Table A 5.1) shows that operating permission was given to 1,38,538 industrial units under Water and Air Acts, and about 5367 industrial units had submitted their environmental statements. The effectiveness of monitoring and enforcement of environmental regulation depends on the compliance of industries with the standards. Table A 5.2 shows that out of 1551 polluting industrial units belonging to 17 water polluting industries in the country, 1178 had adequate facilities to comply with the standards as on 31 March 1995.

The failure of environmental regulation in India may be attributed partly to the type of instruments used (India uses only command and control regulation), and partly to the ineffectiveness of monitoring and enforcement mechanisms. Use of economic instruments, like pollution taxes or marketable pollution permits, and participation of local communities in environmental management can help India achieve the required environmental targets. The Government of India now has several programmes involving NGOs and local communities, for promoting public participation in the environmental management. Recognizing that public participation is pivotal for all environmental programmes, and with a view to strengthening the environmental movement at the grassroots level, MOEF has constituted a Grievance Cell in 1991 and an NGO Cell in 1992. The Grievance Cell attends to the complaints of people regarding environmental problems. The main function of the Cell is to refer the complaints received to the District Magistrates, Municipal Corporations, Pollution Control Boards and other development authorities for redressal. During the year 1994–5, more than 353 complaints have been received and attended to (Table A 5.3). Most of the complaints relate to:

- unauthorized industries located in residential areas discharging harmful gases and hazardous effluents in the immediate neighbourhood;
- environmental degradation due to mismanagement of civic amerities like location of wastewater dump and water logging;
- maintenance of public lands and parks;

• commercial establishments operating illegally in residential buildings, causing nuisance to people living in the immediate vicinity.

A large number of environmental NGOs are now operating in India and playing a catalytic role in the participation of local communities in handling problems, ranging from industrial pollution abatement to preservation of biodiversity and forest resources. The main responsibility of the NGO Cell of MOEF is to attend to visitors and correspondence that seek information on the different schemes and programmes of the MOEF involving NGO participation. Another NGO Cell has also been established by MOEF in the year, 1994–5 at the CPCB to provide training to NGOs and organize mass environmental awareness programmes through them.

Local communities and environmental groups can contest the decisions for giving industrial licenses. Each industrial unit in India has to go through a long process involving the various Ministries of central and state governments for getting industrial licence, environmental clearance and permission to operate. Central and state governments have to give environmental clearance to industrial activities with the objective of achieving efficient use of natural resources by making the polluting firms use better technologies and management practices and incorporate remedial measures at the project formulation stage.

All the projects listed under Schedule-I of the Environmental Impact Assessment (EIA) Notification[8] (Table A 5.4) are required to obtain environmental clearance from the Central Government. Site-specific projects such as mining, pit-head thermal power stations, hydro-power, major irrigation projects, ports and harbours will also have to obtain separate site clearance from the Central Government (or the State Government as the case may be), as specified in the EIA Notification. Over and above this, MOEF notifies certain areas as ecologically sensitive/fragile areas from time to time, and all developmental projects which are to be located in these notified areas need to obtain environmental clearance from the Central Government. Out of 330 projects reviewed for environmental clearance in the calender year 1996, 170 projects were granted environmental clearance. Table A 5.5 provides information about the projects received and projects cleared in different sectors.

[8] MOEF, Government of India, Handbook of Environmental Procedures and Guidelines, 1994.

There are now environmental laws in India, empowering the affected people from environmental pollution to claim compensation from the polluter. These laws have come into force with the National Environmental Tribunal Act, 1995 and the setting up of the National Environment Appellate Authority, 1997. The National Environmental Tribunal Act has come into force from 17 June 1995. It provides for (a) strict liability for damages arising out of any accident that occurs while handling any hazardous substances and (b) establishment of a National Environmental Tribunal for effective and expeditious disposal of cases arising out of such accidents, with a view to giving relief and compensation for damages to persons, property and the environment. This legislation is the first of its kind in the world for providing relief, compensation and restitution to victims of accidents while handling hazardous substances and for environmental damages.

The Government of India has issued the National Environment Appellate Authority Ordinance on 30 January 1997. The Ordinance provides for the establishment of a National Environmental Appellate Authority to hear appeals with respect to restriction of areas in which any industry shall not be carried out or shall be carried out subject to certain safeguards under the Environment (Protection) Act, 1986 and for matters connected therewith or incidental thereto. This is to bring in transparency in the process, accountability and to ensure the smooth and expeditious implementation of developmental schemes and projects. Public hearings have become part of environmental impact assessment. Also, the machinery is being set up for hearing appeals against environmental pollution impact assessment.

## 5.2 ENVIRONMENTAL REGULATION AND LOCAL COMMUNITIES

Compliance of industries to pollution standards can be influenced by firm-specific factors, factors specific to location and the characteristics of local communities. Pargal and Wheeler (1996) have tried to capture the effect of these factors on pollution loads of industrial effluents in Indonesian industries by considering local communities and industries respectively as suppliers and purchasers of environmental services. They found a strong negative relationship between the community characteristics like income and education, and pollution loads of industries. A similar study of Indian industry by Pargal, Mani and Huq (1997) using data for 250 factories belonging to six industries in eight states did not find a significant relationship between the development index of districts (an indicator of affluence of local communities) and BOD loads of factories. In democracies like India, non-economic characteristics of local communities like, their activeness in politics, as reflected in their participation rate in the elections to Parliament, state assemblies and local municipalities can also be important determinants of levels of compliance of industries to pollution abatement standards, along with economic characteristics like income and education levels.

To test the hypothesis *whether both economic and non-economic characteristics of local communities matter in ensuring certain quality of local environmental resources in India*, data for a sample of 100 factories belonging to 11 highly water polluting industries in 13 states of India was collected in a recent survey[9] (Table 5.1). The survey provides data about BOD influent and effluent loads and concentrations, output, age, capital stock, wages and salaries, and capital and operating costs of water pollution abatement for a sample of 100 factories. Using this data of firm-specific factors and Annual Survey of Industries (ASI) data, firm-specific prices of capital and labour inputs are estimated.

The data about region and community-specific factors like population density and development indices of factory locations are obtained from reports of Centre for Monitoring the Indian Economy (CMIE),[10] while the data about the participation rates of

[9] Murty, M. N. et al. (1995), 'IDRC (International Development Research Centre) Project Survey of Water Polluting Industries in India', Institute of Economic Growth, Delhi.

[10] This result is different from the result of a recent World Bank study (Pargal, Mani and Huq, 1997) for the Indian industry which uses data of 250 factories belonging to six industry groups. The current study differs from the above-mentioned study in the following ways:

(i) the sample of factories for the current study is drawn from 11 highly water-polluting industry groups.

(ii) the measurement of pollution of a factory is taken as the ratio of BOD effluent and influent concentrations (BODmg/l).

(iii) political participation of local communities, as given by their rate of participation in the elections, and the number of environment NGOs present in the locality are taken as explanatory variables.

Table 5.1

Number of Factories—State-wise and Industry-wise

| Industry | Fertilizer | Sugar | Cement | Distillery | Chemicals | Refining | Tanning | Iron & Steel | Paper | Drugs | Misc. | Total |
|---|---|---|---|---|---|---|---|---|---|---|---|---|
| Tamil Nadu | 1 | 4 | 2 | 1 | 2 | 5 | 6 | | | 2 | | 23 |
| Kerala | 1 | | | | 1 | 1 | | | 2 | | | 5 |
| West Bengal | 1 | | | 2 | 5 | | | 1 | 2 | 1 | | 12 |
| Andhra Pradesh | | | | 4 | | | 2 | | 2 | 1 | | 9 |
| Haryana | | 2 | | | | | 1 | | 2 | | | 5 |
| Punjab | | | | | | | | | 2 | | | 2 |
| Gujarat | 1 | | | | 8 | 1 | | | | 1 | 5 | 16 |
| Karnataka | | 7 | | 2 | 1 | | | | 3 | 1 | | 14 |
| Maharashtra | | 2 | | | | | | | 1 | | | 3 |
| Orissa | | | | | | | | | 3 | | | 3 |
| Bihar | | 1 | | 1 | | 1 | 1 | 1 | | | | 5 |
| Uttar Pradesh | | | | | | | | | 1 | | 1 | 2 |
| Madhya Pradesh | | | | | | | | | 1 | | | 1 |
| Total | 4 | 16 | 2 | 10 | 17 | 8 | 10 | 2 | 19 | 6 | 6 | 100 |

*Source:* Survey data, 1995–6.

local communities in the recent elections to the Indian Parliament are obtained from the Office of Election Commissioner, Government of India. Table 5.2 reports the summary statistics of various variables for which data are collected.

BOD is the most commonly used index of water pollution. It is normally expressed in milligrams per litre (mg/1) of residual water from industries and households and may be regarded as BOD concentration. The total load of BOD from a factory can be arrived at by multiplying BOD concentration by the volume of residual water. The data from the survey mentioned above, gives information about the average BOD concentrations in influents and effluents from the factories in the sample during the year 1994. The ratio of effluent concentration to influent concentration of BOD (BODR) can be taken as an index of reduction of pollution due to efforts made by the factory. The lower the ratio, the higher are the factory's efforts to reduce pollution.

BODR is modelled as a function of firm-specific variables like output (OP), age (AGE), and input prices (price of capital PK, and price of labour, PL); location-specific variables like population density (PD), per capita state domestic product (PSDP), number of factories (NOF), number of non-governmental organizations (NG) and community-specific variables like development indices (RID), and rate of participation in the last parliament elections (PV). The explanatory variables in the model may be also classified as economic and non-economic factors, the latter consisting of PV and NG. The specification of the function to be estimated is given as follows:

$$\text{BODR} = \text{F}(\text{OP, AGE, PK, PL, PD, PSDP, RID, PV, NG}, D_1, D_2, D_3, \text{U}) \tag{5.1}$$

where $D_1$, $D_2$, and $D_3$ are intercept dummies respectively for the industries like chemicals, including drugs and fertilizers, sugar and distilleries and paper and pulp and U is the disturbance term.

Two models are estimated, one with all the explanatory variables in equation (5.1) (general model) and another without some plant-specific variables like age and industry dummies (restricted model). The results given in Table 5.3 show that there is a strong negative relationship between the affluence of local communities as indicated by the relative development index of a district in which the factory is located, and the rate of BOD reduction obtained by the factories.

Table 5.2
Summary Statistics

| Variable | Unit | Mean | Standard Deviation | Minimum | Maximum |
|---|---|---|---|---|---|
| BOD Ratio (Effluent/Influent) | – | 0.1858 | 0.2402 | 0.0005 | 1.00 |
| Output of the plant | Rupees Million | 1,932.40 | 5,220.60 | 10.31 | 38,226.40 |
| Age of the Plant | Years | 29.32 | 19.74 | 5.00 | 117.00 |
| Price of Labour | Rupees/person/year | 53,100 | 34,200 | 10,000 | 2,60,000 |
| Price of Capital | Rupee per unit of capital | 0.2496 | 0.1065 | 0.04 | 0.46 |
| Number of Factories | Number per state | 10,545 | 5,867 | 1,611 | 18,710 |
| Per capita State Domestic Product | Rupees | 7,483.20 | 18,276.90 | 3,356 | 12,724 |
| Number of NGOs in a District | Number | 8.87 | 11.14 | 1 | 38 |
| Relative Development Index | | 123.74 | 69.39 | 39 | 493 |
| Population Density | per sq. km | 1,000.10 | 3,308.50 | 111 | 23,782 |
| Voting Electorate per constituency | percentage | 60.16 | 15.69 | 31.31 | 86.55 |

In the restricted model, the co-efficient of RID is significant at 1 per cent level, while in the general model, it is significant at 10 per cent level, confirming the hypothesis that the affluence of local communities matters for the compliance of industries with pollution abatement. The elasticities of rate of reduction of BOD(BODR) with respect to relative index of development (RID) are –2.609 and –1.179 respectively for restricted and general models. The restricted model explains that for every one per cent increase in RID, there is 2.609 per cent fall in BODR.

Table 5.3
Regression Results

| Dependent Variable: Log BOD ratio | | | | (N=100) |
|---|---|---|---|---|
| | Model I | | Model II | |
| $R^2$ | 0.3122 | | 0.1605 | |
| Variables | Co-efficient | t-Statistic | Co-efficient | t-Statistic |
| Intercept | –12.03 | –1.74 | –11.93 | –1.70 |
| *Firm-Specific Variables* | | | | |
| Log (output) | 0.11 | 0.75 | 0.25 | 1.85* |
| Log (age) | 0.07 | 0.26 | | |
| Log (wage rate) | 0.82 | –2.10** | –0.85 | –2.13** |
| Log (price of capital) | –1.05 | –3.23*** | –0.46 | –1.45 |
| *Location & Community-Specific Variables* | | | | |
| Log (No. of factories) | –0.06 | –0.17 | –0.43 | –1.24 |
| Log (No. of NGOs) | –0.00 | –0.02 | 0.19 | 1.05 |
| Log (Population density) | 0.37 | 1.27 | 0.67 | 2.27** |
| Log (per capita state domestic product) | 1.27 | 1.32 | 2.07 | 2.14** |
| Log (relative index of development) | –1.18 | –1.85* | –1.69 | –2.61*** |
| Log (Percentage of votes polled) | –0.76 | –1.17 | –1.58 | –2.43** |
| *Dummy Variables* | | | | |
| DI (chemicals including drugs & fertilizer) | 0.93 | 2.10** | | |
| D2 (sugar & distilleries) | –1.34 | –2.55** | | |
| D3 (paper and pulp) | 0.26 | 0.48 | | |

*Note:* *** indicates significance at 1 per cent level; ** indicates significance at 5 per cent level; * indicates significance at 10 per cent level

The results of the restricted model show that there is also strong negative relationship between political participation of local communities as indicated by their rate of participation in the recent elections to Indian Parliament (PV) and the rate of fall in BOD(BODR), with the co-efficient of PV significant at 5 per cent level. The elasticity of BODR with respect to PV is –2.433 implying that for every one per cent increase in political participation of local communities, there is a fall of 2.433 per cent in BODR, supporting the hypothesis that the political activity of local communities matters in increasing the compliance of industries to pollution abatement. In the case of the general model, the co-efficient of PV is not significant even at 10 per cent level, though it is negative as expected.

In the general model, there is a strong significant relationship between BODR and firm-specific variables like input prices and industry dummies. The co-efficients of prices of labour and capital are negative and significant, at 5 and 1 per cent levels respectively. The co-efficients of industry dummies for chemicals and sugar and distilleries are significant at 5 per cent level, the co-efficient of the latter being negative. These results imply that the compliance of sugar and distillery industries with pollution abatement requirements is relatively high, while the compliance of chemical industries is relatively less in comparison to other water polluting industries in India.

## 5.3 CONCLUSIONS

Local communities can influence pollution abatement decisions of industries, both with and without formal environmental regulations. There is now enough evidence from the developing countries, about the informal environmental regulation by the local communities when formal regulation is absent or ineffective. In the developed countries even with formal regulation, there is now a growing tendency of involving local communities and environmental groups in all activities of environmental management including design, monitoring and implementation of environmental laws. This phenomenon is viewed as the privatization of environmental management, with laws permitting the local environmental groups and industries to contest government regulations.

The compliance of industries with pollution abatement depends upon firm-specific and location- and community-specific factors.

Recent studies of some developing countries show that location- and community-specific factors are important in facilitating informal regulation when formal regulation is ineffective. This study finds this to be true in the case of India as well. Results show that in India, the affluence of local communities as reflected in the higher relative development index of localities to which they belong, and the level of political activity as shown by their rate of participation in the elections to parliament, play important roles in determining the degree of compliance of industries with water pollution abatement standards.

# APPENDIX A 5

Table A 5.1
Consents Granted and Environmental Statements Submitted by Industries

| S.No. | State/Union Territory | Consents Granted under | | | Total Number of Environmental Statements | Total Number of Industries | |
|---|---|---|---|---|---|---|---|
| | | Water Act | Air Act | Total | | Large & Medium | Small |
| 1. | Assam | 535 | 126 | 661 | 10 | 377 | 10,000 |
| 2. | Andhra Pradesh | 356 | 299 | 655 | 83 | 828 | 91,162 |
| 3. | Arunachal Pradesh | 5 | 5 | 10 | 0 | 10 | 376 |
| 4. | Bihar | 504 | 1,226 | 1,730 | 237 | 389 | 59,886 |
| 5. | Gujarat | 5,072 | 3,770 | 8,842 | 473 | 924 | 5,853 |
| 6. | Goa | 661 | 77 | 738 | 19 | – | 21 |
| 7. | Haryana | 1,500 | 948 | 2,448 | 94 | 183 | 61,229 |
| 8. | Himachal Pradesh | 481 | 466 | 947 | 57 | 69 | 9,336 |
| 9. | Jammu & Kashmir | 102 | – | 102 | – | – | – |
| 10. | Karnataka | 1,278 | 1,145 | 2,423 | 288 | 1,206 | 4,935 |
| 11. | Kerala | 908 | 187 | 1,095 | 93 | – | 315 |
| 12. | Maharashtra | 56,907 | 17,592 | 74,499 | 2,158 | 4,000 | 1,28,700 |
| 13. | Madhya Pradesh | 4,008 | 3,705 | 7,713 | 210 | 744 | 3,52,749 |
| 14. | Manipur | 4 | 4 | 8 | – | 8 | 281 |
| 15. | Meghalaya | 57 | 59 | 116 | 5 | 7 | 1,006 |
| 16. | Mizoram | – | – | – | – | – | 2,145 |
| 17. | Nagaland | – | – | – | – | – | 509 |

*(Cont.)*

| S.No. | State/Union Territory | Consents Granted under | | | Total Number of Environmental Statements | Total Number of Industries | |
|---|---|---|---|---|---|---|---|
| | | Water Act | Air Act | Total | | Large & Medium | Small |
| 18. | Orissa | 447 | 359 | 806 | 136 | 217 | 16,061 |
| 19. | Punjab | 5,804 | 4,089 | 9,893 | 100 | 425 | 95,000 |
| 20. | Rajasthan | 362 | 260 | 622 | 218 | 250 | 56,761 |
| 21. | Tamil Nadu | 12,044 | 8,548 | 20,592 | 297 | 2,527 | 12,203 |
| 22. | Tripura | 529 | 529 | 1,058 | 1 | 3,727 | – |
| 23. | Uttar Pradesh | 552 | 519 | 1,071 | 404 | 464 | 1,45,797 |
| 24. | West Bengal | 445 | 137 | 582 | 250 | – | 1,21,656 |
| 25. | Andaman & Nicobar | – | – | – | – | – | – |
| 26. | Chandigarh | 300 | 100 | 400 | 8 | 20 | 400 |
| 27. | Daman, Diu & D.N. Haveli | 1,087 | 68 | 1,155 | 193 | – | – |
| 28. | Delhi | – | – | – | – | – | – |
| 29. | Lakshadweep | – | – | – | – | – | – |
| 30. | Pondicherry | 227 | 195 | 422 | 33 | 83 | 4,784 |
| 31. | Sikkim | – | – | – | – | 9 | 270 |
| | Total | 94,175 | 44,413 | 1,38,538 | 5,367 | 13,076 | 11,99,159 |

*Source:* CPCB, Annual Report, 1994–5.

Table A 5.2

Statewise Summary Status of Pollution Control in 17 Categories of Industries

As on March 31, 1995

| S. No. | State/Union Territory | Total Number of Units Belonging to 17 Categories | Status (Number of Units) | | | |
|---|---|---|---|---|---|---|
| | | | Closed | Able to Comply with Standards | Unable to Comply with Standards | |
| | | | | | Pre-1981 | Post-1981 |
| 1. | Andhra Pradesh | 173 | 34 | 107 | 22 | 10 |
| 2. | Arunachal Pradesh | 0 | 0 | 0 | 0 | 0 |
| 3. | Assam | 15 | 0 | 10 | 5 | 0 |
| 4. | Bihar | 62 | 14 | 37 | 10 | 0 |
| 5. | Delhi | 5 | 0 | 2 | 2 | 1 |
| 6. | Goa | 6 | 0 | 6 | 0 | 0 |
| 7. | Gujarat | 177 | 2 | 167 | 5 | 3 |
| 8. | Haryana | 43 | 3 | 33 | 6 | 1 |
| 9. | Himachal Pradesh | 9 | 0 | 9 | 0 | 0 |
| 10. | Jammu & Kashmir | 8 | 3 | 1 | 1 | 3 |
| 11. | Karnataka | 85 | 6 | 58 | 19 | 2 |
| 12. | Kerala | 28 | 4 | 20 | 4 | 0 |
| 13. | Madhya Pradesh | 78 | 2 | 55 | 13 | 8 |
| 14. | Maharashtra | 335 | 14 | 293 | 28 | 0 |
| 15. | Manipur | 0 | 0 | 0 | 0 | 0 |
| 16. | Meghalaya | 1 | 0 | 0 | 1 | 0 |

(*Cont.*)

| S. No. | State/Union Territory | Total Number of Units Belonging to 17 Categories | Status (Number of Units) | | | |
|---|---|---|---|---|---|---|
| | | | Closed | Able to Comply with Standards | Unable to Comply with Standards | |
| | | | | | Pre-1981 | Post-1981 |
| 17. | Mizoram | 0 | 0 | 0 | 0 | 0 |
| 18. | Nagaland | 0 | 0 | 0 | 0 | 0 |
| 19. | Orissa | 23 | 0 | 13 | 9 | 1 |
| 20. | Punjab | 45 | 2 | 32 | 5 | 6 |
| 21. | Pondicherry | 6 | 0 | 2 | 3 | 1 |
| 22. | Rajasthan | 49 | 5 | 42 | 2 | 0 |
| 23. | Sikkim | 1 | 0 | 0 | 1 | 0 |
| 24. | Tamil Nadu | 119 | 2 | 109 | 4 | 4 |
| 25. | Tripura | 0 | 0 | 0 | 0 | 0 |
| 26. | Uttar Pradesh | 224 | 16 | 168 | 38 | 2 |
| 27. | West Bengal | 58 | 10 | 21 | 26 | 1 |
| 28. | Andaman & Nicobar | 0 | 0 | 0 | 0 | 0 |
| 29. | Chandigarh | 1 | 0 | 1 | 0 | 0 |
| 30. | Daman & Diu | 0 | 0 | 0 | 0 | 0 |
| 31. | Dadra & Nagar Haveli | 0 | 0 | 0 | 0 | 0 |
| 32. | Lakshadeep | 0 | 0 | 0 | 0 | 0 |
| | Total | 1,551 | 121 | 1,178 | 207 | 45 |

*Source:* CPCB, Annual Report, 1994–5.

Table A 5.3
Status of Public Complaints (Statewise) during 1994–5

| State/Union Territory | Number of Complaints Received | Number of Complaints Disposed of |
|---|---|---|
| Andhra Pradesh | 3 | Nil |
| Assam | 1 | Nil |
| Bihar | 10 | Nil |
| Goa | 5 | 2 |
| Gujarat | 11 | 2 |
| Haryana | 10 | Nil |
| Himachal Pradesh | 1 | Nil |
| Jammu & Kashmir | 11 | Nil |
| Karnataka | 12 | Nil |
| Kerala | 13 | Nil |
| Madhya Pradesh | 10 | 3 |
| Maharashtra | 28 | 2 |
| Manipur | 1 | Nil |
| Orissa | 15 | 1 |
| Punjab | 10 | Nil |
| Rajasthan | 10 | 1 |
| Uttar Pradesh | 33 | Nil |
| West Bengal | 38 | 1 |
| Delhi | 130 | 9 |
| Daman, Diu & Dadar Nagar Haveli | 1 | Nil |
| Total | 353 | 21 |

*Source:* CPCB, Annual Report, 1994–5.

Table A 5.4
Industries for which Licensing will be Compulsory

1. Coal and Lignite
2. Petroleum (other than crude) and its distillation products
3. Distillation and brewing of alcoholic drinks
4. Sugar
5. Animal fats and oils
6. Cigars and cigarettes of tobacco and manufactured tobacco substitutes
7. Asbestos and asbestos-based products
8. Plywood, decorative veneers, and other wood-based products such as particle board, medium-density fibre board, block board
9. Raw hides and skins, leather, chamois leather and patent leather
10. Tanned or dressed furskins

(*Cont.*)

11. Motor cars
12. Paper and Newsprint except bagasse-based units
13. Electronic aerospace and defence equipment: all types
14. Industrial explosives, including detonating fuse, safety fuse, gun powder, nitrocellulose and matches
15. Hazardous chemicals
16. Drugs and Pharmaceuticals (according to drug policy)
17. Entertainment Electronics (VCRs, Colour TVs, CD Players, Tape Recorders)
18. White goods (Domestic Refrigerators, Domestic Dishwashing Machine, Programmable Domestic Washing Machines, Microwave ovens, Airconditioners).

*Source:* GOI, Ministry of Industry, Handbook of Industrial Statistics, 1995.

Table A 5.5
Status of Development Projects Gone to MOEF for Clearance

| Sector | Projects Received | Projects Cleared |
|---|---|---|
| 1. Mining Projects | 78 | 54 |
| 2. Industrial Projects | 144 | 54 |
| 3. Power Projects | 66 | 32 |
| 4. Infrastructure | 42 | 30 |
| Total | 330 | 170 |

*Source:* MOEF Annual Report, 1996–7.

# 6. Collective Action for Pollution Control in Practice: Experiences of Some Industrial Estates in South India[1]

M. N. MURTY and U. R. PRASAD

The fixing and implementation of pollution standards in India supports the Coasean bargaining framework. National pollution standards in India are fixed through a political process involving scientists, politicians and affected people. Implementation of the standards involves collective action of industries, Pollution Control Boards, NGOs, and the affected people. State Pollution Control Boards can change the national standards by making them more stringent to suit the local ambient environmental quality, but they cannot relax them. Hence, the final outcome of the local standards depends on the bargaining between State Pollution Control Boards, NGOs and the affected people. Thus national environmental standards could be different from local or state standards in India, which is actually the case in many states. In the Coasean bargaining framework, both the fixing and the implementation of standards are part of the contract. The empirical evidence in India suggests the presence of many such contracts in industrial water pollution abatement.

## 6.1 EXPERIENCES OF SOME INDUSTRIAL ESTATES IN INDIA

The survey of Jeedimetla, Bollaram and Patancheru industrial estates in Andhra Pradesh and Vaniambadi industrial estate in Tamil Nadu;

[1] This chapter forms part of research for the project 'Fiscal Instruments for Water Pollution Abatement in India'. We are grateful to IDRC, Canada for financial assistance.

and Nandesari industrial estate in Gujarat, (Misra, 1996) provides insights into the empirical aspects of the economic and non-economic processes leading to collective action in industrial pollution abatement. There are three processes involved in the collective action for controlling water pollution in an industrial estate (Misra and Murty, 1995a). These are (a) collective action of affected parties, (b) collective action of factories and (c) the bargaining between a coalition of affected people and a coalition of factories. The organization of affected people as a coalition depends upon their perception of damages. If damages from pollution are fully perceived by the public and they are substantive enough to justify transaction costs of forming a coalition and bargaining, collective action follows. The transaction costs of collective action of affected public are in the form of costs of dissemination of information about damages among the people, the costs of public litigation, legal action against factories and government, etc. Given the deterrent of collective action by the affected people, the factories in an industrial estate have to take recourse to pollution abatement methods to improve the effluent quality as per the prescribed standards. The available pollution abatement technologies have provided the factories with a broad spectrum of technological choices, ranging from changes in production process and end of pipe treatment methods within the factory, to CETP for a coalition of factories. The presence of scale economies in water pollution abatement has provided incentives to factories to form a coalition for the adoption of CETP technology (Misra and Murty, 1995a). Also, co-opeartive efforts are needed to mobilize resources for R&D to develop in plant abatement technologies which are suitable for the small-scale enterprises in an industrial estate. Therefore, collective action by factories for water pollution abatement is technology-driven. However, there are other factors like the problems of dealing with a coalition of affected parties, government and financial institutions, that have driven the factories to resort to collective action. Given the coalitions of affected parties and factories, bargaining between them has finally produced the end result of collective action, that is, the realization of the prescribed effluent standards by the industrial estate.

It is found that various agents like people affected by water pollution, elected representatives (members of Parliament, State Assemblies, Municipal Committees), industries, NGOs and government are involved in the processes leading to the establishment of CETPs.

The case studies show that there is a complementarity between collective action for the preservation of 'water quality', and the interests of local communities and the government. They highlight a 'catalytic' role of government instead of a 'coercive' role in bringing about water pollution abatement. The causes for the emergence of collective action may be summed up as follows:

• Mutually supporting interests of elected representatives and people affected by pollution. Affected people need the support of elected representatives to deal with industry and bureaucracy, and the elected representatives need the support of affected parties to get re-elected.

• Catalytic role of NGOs and the bureaucracy

• Presence of a property rights regime which supports the 'polluter pays' principle.

• Presence of governmental regulations implemented through State Pollution Control Boards.

• Presence of scale economies that are characteristic of water pollution abatement.

Processes underlying collective action in the industrial estates considered, have historically started with one shot 'Prisoners' Dilemma' type of game among the affected people, resulting in a situation of non-cooperation for controlling water pollution. In this case, the Prisoners' Dilemma type of problem might have arisen for sharing the transaction costs, the costs of educating the people about the damages from water pollution, the costs of lobbying for enrolling the affected people in the association, and the costs of legal action against the industry and government. Given that the output of collective action is the power of local communities to deal with industry and government through legal action or other means for controlling water pollution, each affected person would like to enjoy the fruits of collective action without actually contributing to its cost. In many cases, the stalemate of non-cooperation was broken by an outside agency like an NGO, by educating the people about the damages from water pollution and benefits that can be obtained from co-operation. Also, co-operation or collective action is possible if the affected people are historically placed in a situation of playing an infinite repetitive Prisoners' Dilemma game. In breaking a stalemate of non-cooperation of affected people in water pollution abatement, NGOs are found to play an important role in the industrial estate areas of Andhra Pradesh, Tamil Nadu and Gujarat. In the Jeedimatla and Nandesari industrial estate areas, NGOs called Jagruti and Soclean have respectively played important roles in this context.

Due to the limitations of collective action at the local level, the government plays a critical role through the following activities:

• Processing information on 'water quality'

• Providing technical and financial assistance in setting up the CETP, an effluent channel, etc.

• Providing economic incentives through penalties and threats of closure.

• Providing a legal framework (polluter pays principle) to legally enforce people's right to call upon the State to act as the enforcer of last resort, (Wade, 1987).

• Providing a monitoring framework.

In the following sections of this chapter, an attempt is made to explain the economic and non-economic processes leading to collective action in water pollution abatement in some industrial estates in South India. The industrial estates studied in this paper include Patancheru, Bollaram and Jeedimetla in Andhra Pradesh and Vaniambadi and other industrial estates in Tamil Nadu. Patancheru with 276 industrial units, Bollaram with 40 industrial units and Jeedimetla with 600 industrial units are situated within a radius of 35 kms from Hyderabad. Vaniambadi and other industrial estates are located in the North Arcot district of Tamil Nadu.

## 6.2 DAMAGES FROM WATER POLLUTION IN THE INDUSTRIAL ESTATES

People living in and around the industrial estates mentioned above have suffered damages in terms of losses of crops, cattle and agricultural equipment; contamination of drinking water; diseases and deaths due to water pollution. In Jeedimetla and Patancheru industrial estates of Andhra Pradesh, more than 7000 acres of productive farmland have been contaminated with toxic chemical run-off, while more than 5,00,000 people have been adversely affected, either through diseases arising out of water pollution, or loss of traditional livelihood. In Jeedimetla industrial estate, within a 10 km radius of Ayodhyanagar, all the drinking water and agricultural wells are contaminated with toxic chemicals. The effect on traditional livelihood is equally devastating. Toxic factory emissions have burnt out the tops of toddy palms within a radious of one kilometre, depriving toddy tappers of their only source of income (Sarangi and Cohen, 1995). A study prepared by NEERI (1991) has identified the extent of losses in agricultural production and cattle losses in the villages in Patancheru and Bollaram industrial areas. As many as 629 farmers are affected with the area of damaged crop land amounting to 1602 hectares in 15 villages. The value of crop loss is estimated at Rs 517 lakhs. Also, as many as 1323 cattle deaths are reported in seven villages in this area due to industrial pollution. About 150 pumpsets are reported to have rusted and become obsolete due to water pollution in this area. About 59 fishermen and 40 washermen have had to change their professions due to pollution of ponds and streams. Many agriculturists have turned industrial workers due to water pollution. Tables A 6.1 and A 6.2 in the Appendix to this chapter provide information about the extent of land, crop and cattle loss. An independent survey of a sample of 62 households (the results of which are reported in Table A 6.3) conducted by our research team in February 1995 shows that polluted land per family is 2.7 acres. The annual per household and per capita income of affected households in the area covering the industrial estates of Patancheru, Bollaram and Jeedimetla are estimated as Rs 10,992 and Rs 1,615 respectively. Table A 6.3 provides this information. Groundwater pollution has affected drinking water in most of the villages. During recent years, drinking water has been supplied to some of the villages from the Manjeera river project on the payment of Rs 20 per month by each household. Many other villages are currently using contaminated groundwater.

A sample survey of Medak district conducted by the office of the District Medical Officer clearly provides evidence of a high incidence of diseases from water and air pollution in the villages affected by industrial estates, in comparison to the control villages which are not affected by pollution. According to this survey, the incidence of disease is 21 per cent higher in the pollution-affected villages in comparison to the villages not affected by pollution. Table A 6.4 provides this information. A survey conducted by our project research team during February 1995 shows that as many as 73 people are affected by various industrial pollution-related diseases in 62 households surveyed in Patancheru, Bollaram and Jeedimetla areas. On an average, an affected person incurred annual medical expenses amounting to Rs 1290 with 119 days of hospitalization in a year. Table A 6.5 provides this information. Dr Krishna Rao, a doctor belonging to an NGO in Patancheru, has reported to our project survey team that a large number of people in the Patancheru industrial area are currently suffering from the following diseases: skin allergies, loss of vision/defective vision, nasal disorders, respiratory diseases, chemical diarrhoea, stomach cancer, dyspepsia, liver disorders due to heavy metals in water, kidney failure, incoherence, nerves epilepsy, mental disorder, paraplizia (bone disease), menstrual disorders, foetal malformation and joint pains. According to Dr Krishna Rao, there is one village in the area where all women have stopped menstruating due to exposure to toxins. In Sultanpur village, 10 families have members who have become incurably paralyzed after drinking contaminated water (Sarangi and Cohen, 1995). Table A 6.6 provides information about villages affected by pollution and compensation paid towards damages.

Tanneries in Ambur, Vaniambadi, Erode, Ranipet, Dindigal and Pallavaram industrial estates of Tamil Nadu have been discharging untreated effluent into the land, tanks and rivers. The entire surface and subsoil water of Palar river, the main source of water supply to the people of Ambur, Vaniambadi, Pernambut, Vellore, Arcot, Ranipet and Kaveripakkam, has been polluted by salts and toxic substances from the tanneries. The agricultural land in the Palar river basin has become unproductive resulting in crop losses, and erosion in property and land values. A preliminary survey made by the Tamil Nadu Agricultural University Research Centre at Vellore on the effect of pollution caused by tanneries in North Arcot district shows that out of one lakh hectares of agricultural land in the 1000 sq km tannery belt running from Vaniambadi to Walajpet, nearly 35,000 hectares of agricultural land have been rendered either partially or totally unfit for cultivation. Also, the area under paddy is drastically reduced by 40 per cent in the tannery belt. According to another survey conducted by Peace Trust, an NGO covering 13 villages of Dindigal and Peddiar Chatram panchayat unions, 350 wells out of a total of 467 used for both drinking and irrigation purposes have been polluted. Women and children have to walk miles to get drinking water. Wells of 1090 houses out of 1298 surveyed, have been damaged due to high salt content. Also, some instances of people deserting the villages due to the pollution are recorded in the tannery belt of Tamil Nadu. For instance, T. Pudur village has only seven out of the original 30 families. The stunted coconut trees, shorn of all foliage and barren lands, are testimony to the devastation caused by pollution in this area.

## 6.3 COLLECTIVE ACTION BY AFFECTED PEOPLE

The process underlying collective action for water pollution abatement has historically started with a situation of non-cooperation among the affected people in the industrial estates surveyed. In breaking a stalemate of non-cooperation, NGOs are found to play an important role in the industrial estate areas of Andhra Pradesh and Tamil Nadu. In the Jeedimatla industrial estate area, an NGO called Jagruti has played an important role in this context. The following are some of Jagruti's activities.

• For increasing awareness among people it has started a quarterly journal, the first environmental news magazine in Telugu, 'Mana Paryavaranam' in 1991. Since then, issues of industrial pollution at Jeedimetla have been highlighted in the journal.

• To create awareness at the grassroots level, Jagruti has been conducting classes for school children on environment, particularly on environmental protection through Eco Clubs. About 30 Eco Clubs are already functioning in the villages Illendu, Kottaguda, Manuguru, Piduguralla and Nagarjuna Sagar in Ranga Reddy, Guntur and Prakasam districts of Andhra Pradesh. Also, it has been conducting Nature Orientation Camps for students to improve their understanding of environmental protection and pollution abatement issues.

• On 19 July 1993, it conducted a seminar on 'Impact of Environmental Degradation on People' which was attended by the speaker of Andhra Pradesh State Legislative Assembly and some MLAs.

• It is going to translate some of the relevant books on environment into Telegu and keep them in the libraries which are to be established in the villages of Andhra Pradesh.

• It has conducted a health survey in the Jeedimetla industrial area to know the effects of water and air pollution on human life. It was found that out of 2000 women surveyed, there were 560 abortions and high rates of skin diseases, loss of vision, respiratory diseases, cancer, liver disorders, menstrual disorders and foetal malformation.

It has been conducting classes, especially for women, on environment and its impact on human life and suggesting some precautionary measures against pollution.

• To control the illegal activities of industry (some of the industries discharge their effluent directly into the open drains) it has formed a band of workers called 'Green Army'. Officers of the Pollution Control Board and member of a safety committee from HMT have also joined this army.

• To work as a receiver agency at the State level, it is going to extend branches of its organization throughout Andhra Pradesh.

Jagruti has started Jeedimetla Kalushya Porata Samiti which is an organization to protest against pollution and through which local people participate in protests and dharnas against industrial pollution and pressurize the government and industry to take corrective measures. Local politicians, realizing the threat of not getting re-elected, joined hands with local people for representing the pollution problem to the state government. As a result, a unit of Andhra Pradesh State Pollution Control Board has been established at Jeedimetla and industries in Jeedimetla were made to have a CETP in 1989.

A local activist and an environmentalist has started 'Patancheru Anti-Pollution Committee' to deal with the problem of water pollution in Patancheru and Bollaram industrial areas. In Patancheru, local opposition started in 1986 when 3000 farmers marched to the Chief Minister's office after suffering large-scale crop losses due to contamination of local wells and the Nakkavagu river. In 1989, 5000 people held a demonstration before the state assembly, demanding an end to industrial pollution. In the same year, farmers blockaded the highway running through Patancheru for two days with their bullock carts. Some time after that, seven members of the 'Patancheru Anti-Pollution Committee' were arrested and imprisoned by the police on trumped-up murder charges. The affected people have made a number of representations to the government and industry without success. In private conversations with the project team during the survey, some factory managers reported that once the affected people had blocked the roads leading to the factories and had threatened to cause physical harm if they tried to open the factories. Notwithstanding the threat, the affected people had also filed court cases against the factories by jointly sharing the cost of public litigation. The villagers reported to the project team during the survey that they have contributed Rs 100 per acre or Rs 200 per head towards the cost of court cases. In one such case: Supreme Court WP(C)No.1050/90, Indian Council for Enviro-Legal Action and Others versus Union of India, the main issues involved in the petition were:

• Providing water supply to the affected villages.

• Providing medical facilities and free treatment to affected people and their livestock by deputing teams of doctors from different specialities (including veterinarians) to pollution-affected areas.

• Preventing pollution from further spreading from surface and groundwater sources and by taking immediate corrective steps.

• To immediately close down hazardous industries that cause air and water pollution.

• Payment of adequate compensation to all the victims of pollution by the government and recovering the cost from the polluting industries/polluters.

• To take serious action against industries flouting environmental laws.

• To constitute a team of environmental experts and economists for an on-the-spot survey and study of water and air pollution and losses incurred on account of pollution, and furnish a report to the Supreme Court.

In response to this court case, the Supreme Court of India has ordered (Supreme Court Order on 11.12.1989 in W.P. No.C967/89) the constitution of an expert committee for undertaking a study of pollution problems in the affected areas. The expert committee has submitted its recommendations with respect to the following issues in the writ petition:

• Providing water supply by tankers to the affected villages mentioned in the writ petition on a temporary basis till a permanent scheme is provided.

• Addressing health aspects in the affected villages referred to in the writ petition.

• Payment of compensation.

• Environmental awareness and training.

• Common Effluent Treatment Plants.

• Monitoring.

This legal action through the collective efforts of affected people has ultimately forced the factories in the industrial estates of

Patancheru and Bollaram to have CETPs. In the case of tanneries in North Arcot district of Tamil Nadu, a series of dharnas, demonstrations and joint action have been mobilized by the affected people since 1985. A protest march was carried out in which 20,000 affected people participated. Women carried the pots of polluted drinking water and broke them before the municipal authorities demanding clean water. The Tamil Nadu Pollution Control Board, which was set up in 1982 could not do much, except to give notices to the tanneries to stop pollution. The time limits given to the tanneries to install an effluent treatment plant have been extended fifteen times.

The government, which is supposed to take care of affected people, has failed in its duty in several ways. It has encouraged ill-planned industries from the beginning without taking into consideration the environmental impact. Tanneries in Ambur, Vaniambadi, Erode, Ranipet, Dindigal, and Pallavaram areas are functioning in blatant disregard for legislations pertaining to water and air pollution abatement. The Vellore Citizens Welfare Forum has filed a writ petition in the Supreme Court of India against the Union of India, State of Tamil Nadu and Tamil Nadu Pollution Control Board about the control of pollution from tanneries in the state. This is a writ petition under Article 32 of the Constitution of India for Writ of Mandamus or an appropriate writ order or direction, for the protection of thousands of innocent lives, suffering from serious air and water pollution being caused by enormous discharge of untreated effluents, by the tanneries into agricultural fields, roadsides, waterways and open lands in North Arcot district of Tamil Nadu. The main issues involved in the prayer are as follows:

- To provide clean drinking water to the people living in the villages being affected by pollution in North Arcot district, Tamil Nadu.
- To immediately send a team of medical doctors from different specialities to the pollution-affected areas for providing treatment to the suffering people.
- To take immediate measures to avert pollution from spreading further in the surface and groundwater sources, and periodically collect water samples and inform people about the quality of water, through medical and other means of providing information.
- To immediately close down tanneries having no effluent treatment plant and flouting pollution control measures and environmental laws.

• To immediately pay adequate compensation to victims of pollution, and to those who lost their lives, crops, vegetation, trees, agriculture land, wells, and suffered severe hardship due to the irresponsible and negligent behaviour of polluting tanneries.

• To take serious action against polluting industries violating environmental laws and showing utter disregard for the safety and welfare of the workers working in tanneries and people affected by the air and water pollution.

• Constitute a team of eminent experts and economists for survey and study water and air pollution and losses incurred on account of pollution and submit a report to this Honourable Court.

In response to this collective action of affected people, many industrial estates containing tanneries in Tamil Nadu are now going for CETPs to control water pollution.

## 6.4 COLLECTIVE ACTION OF FACTORIES

In the case of industrial estates with small-scale enterprises, the compulsions arising out of collective action of affected people to take measures to control water pollution and the available technological options for pollution abatement can necessitate collective action by the industries. Due to the presence of economies of scale in water pollution abatement, it may not be economical for each factory to have its own effluent treatment plant. There is a need either for having a CETP for all the factories or for incurring expenditures on research and development to innovate cost-effective, factory-specific pollution abatement technologies for small factories which is possible only through collective action by the factories. Alternatively, collective action may not be needed if waste-treatment facilities are offered to the factories by a private limited company at a price. Misra and Murty (1995a) have discussed the incentives and impediments to collective action for water pollution abatement by the small factories in an industrial estate. They have highlighted the need for government and NGOs playing a catalytic role in the collective action by the factories.

Use of CAC instruments by the government in an environment of non-availability of economically viable technological options for pollution abatement has caused considerable hardships to small-scale enterprises. Given that the government managed public sector has been the fountainhead of industrial development in India during all these years, the government has not made any sincere efforts to promote economically viable pollution abatement technologies for the small-scale enterprises via investments in R&D in the public sector. Some effort has been made only recently, especially with the financial support of international agencies like the World Bank, to promote pollution abatement technologies like the CETP for small-scale enterprises.

In the industrial estates of Andhra Pradesh and Tamil Nadu, collective action for water pollution abatement is technology-driven. Associations of factories are formed for having CETPs and for agreeing on a common set of rules governing CETPs. There are CETPs already functioning in Jeedimetla, Patancheru, Bollaram, Pallavaram and Vaniambadi industrial estates and many are nearing completion in other industrial estates. Tables A 6.7, A 6.8 and A 6.9 provide information about some of these CETPs.

Take for example, collective action by the factories in Jeedimetla industrial estate. The agreement of the association of factories at Jeedimetla clearly states that the factories have examined the possibility of having their industrial effluents treated at its factory site and found it to be economically unviable. In this case, cost considerations have triggered collective action of factories and prompted factories to form an association for floating a company called Jeedimetla Effluent Treatment Limited. Each factory buys shares of the treatment company and enters into an agreement about the set of rules governing the company. Share-holding thus provides assurance to the factories of the treatment plant's services on a mutually agreed cost-sharing arrangement. This allows for sustaining collective action in water pollution abatement in Jeedimetla industrial estate.

The agreement among factories focuses on the following:

A. Quantity and quality standards of effluents to be met by the factories:

1. the capacity of each tanker should be 8000–10,000 litres.
2. pH level of industrial effluents to be between 7 and 7.5.
3. the industrial effluents should not contain heavy metals, cyanides, poison, etc.
4. the customer shall be bound by the chemical analysis of its industrial effluent carried out by the Treatment Company.
5. the customer shall be bound by the calibration taken by the Treatment Company for determining the quantity of the industrial effluent sent for treatment by the customer.

In case the above standards are not met, a charge of Rs 100 per rejected tanker is levied. This ensures pre-treatment by the factory, enabling a particular quality effluent to reach the Treatment Plant. This would allow a homogeneous quality effluent to be treated at the Treatment Plant which could simplify the method of charging the factories for the effluent treatment.

B. Effluent-specific charges on the basis of
   1. the level of COD
   2. the level of pH
   3. the amount of Total Dissolved Solids (TDS)

The pollutant-specific effluent charges levied by the treatment company are given in Table A 6.10.

The agreement also makes sure that a minimum rent of Rs 75 per tanker per day is payed by the factory even when it fails to send any effluent to the treatment plant. This would ensure an even distribution of the treatment plant's services over the month, and avoid congestion caused by accumulated tankers on certain days.

Given the property rights and the environmental law in force, it becomes obligatory on the part of the factories to get their effluents treated. The mutually drawn cost-sharing agreement ensures the sustainability of collective action, enabling the preservation of water quality. Thus, the agreement among the factories is an important tool in bringing about collective action.

## 6.5 CONCLUSION

Experiences of industrial estates surveyed show that collective action is emerging as an alternative institution to economic instruments for industrial water pollution abatement. Collective action of affected households has resulted in threats of closure to factories by blocking roads, and legal action against the government and factories. Given the threat from households, factories are compelled to go for pollution abatement through a contract for having a CETP. The threat from households has also resulted in the government playing a catalytic role in promoting the use of CETP in the industrial estates.

The rising damages from water pollution, efforts of local NGOs in increasing the awareness of households about these damages and realization of local politicians that it is in their interest to fight against pollution have ultimately resulted in the emergence of collective action of households in the vicinity of industrial estates. The growing number of legal cases pertaining to environmental pollution, prompted by the affected local population, and the success of many of these cases is a testimony to the growing importance of collective action as an institution, to deal with environmental pollution in India.

Small-scale enterprises cannot reap the benefits from the fiscal incentives provided by the government for the control of pollution since the accrual of benefits depends upon the installation of pollution control equipment by the factory and small-scale enterprises cannot mobilize the resources to have their effluent treatment plants. Use of CAC instruments has compelled small-scale enterprises to use economically unviable pollution abatement technologies and to face the threat of closure in case of non-compliance with prescribed pollution standards.

# APPENDIX A 6

Table A 6.1
Extent of Damage in Various Villages of Patancheru and Bollaram Industrial Estates

| S. No. | Village | Affected Farmers | Area Damaged (Hectares) | Crop Loss (Rupees Lakhs) |
|---|---|---|---|---|
| 1. | Peddakanjerla | 57 | 100 | 35 |
| 2. | Byathole | 40 | 70 | 20 |
| 3. | Sultanpur & Gandigudem | 70 | 110 | 65 |
| 4. | Chitkul | 18 | 85 | 8 |
| 5. | Bachugudem | 38 | 42 | 18 |
| 6. | Chidruppa | 39 | 85 | 10 |
| 7. | Lakdaram | 28 | 22 | 3 |
| 8. | Inole | 14 | 40 | 26 |
| 9. | Arutla | 62 | 412 | 120 |
| 10. | Ganapathigude | 25 | 40 | 16 |
| 11. | Eddanoor | 14 | 132 | 30 |
| 12. | Ismailkhanpet | 20 | 133 | 70 |
| 13. | Kistareddipet | 150 | 200 | 80 |
| 14. | Masnur | 6 | 8 | 2 |
| 15. | Patanchru | 48 | 123 | 14 |
| | Total | 629 | 1,602 | 517 |

*Source*: NEERI, 1991.

Table A 6.2
Cattle Loss in Various Villages of Patancheru and Bollaram Industrial Estates

| S. No. | Village | Sheep | Ox | Goat | Buffalo | Cow | Total |
|---|---|---|---|---|---|---|---|
| 1. | Peddakanjerla | 24 | 16 | 28 | 122 | 2 | 192 |
| 2. | Sultanpur | – | 22 | – | 54 | 2 | 78 |
| 3. | Chitkul | – | 1 | – | 3 | – | 4 |
| 4. | Lakdaram | – | – | – | 4 | – | 4 |
| 5. | Inole | – | 8 | – | 16 | 1 | 25 |
| 6. | Eddanoor | 555 | 49 | 239 | 99 | 41 | 983 |
| 7. | Pocharam | – | 15 | – | 21 | 1 | 37 |
| | Total | 579 | 111 | 267 | 319 | 47 | 1323 |

*Source:* NEERI, 1991.

Table A 6.3

Income and Other Losses of Affected Persons in Three Industrial Estates of Andhra Pradesh

| S. No. | Village | Number of Families Surveyed | Members of the Families | Total Annual Income | Per Capita Income | Affected Land (Acres) | Annual Crop Loss (Rs) | Annual Cattle Loss | Compensation Received |
|---|---|---|---|---|---|---|---|---|---|
| *Patancheru Area* | | | | | | | | | |
| 1. | Lakdaram | 5 | 43 | 27,000 | 628 | 10.0 | 0 | 0 | 5,600 |
| 2. | Chitkul | 5 | 31 | 67,000 | 2,161 | 27.0 | 0 | 1 | 4,500 |
| 3. | Ganapathtigudem | 5 | 28 | 25,400 | 907 | 28.5 | 0 | 2 | 13,051 |
| 4. | Vadigapalli | 4 | 24 | 65,000 | 2,708 | 4.5 | 2,000 | 0 | 0 |
| 5. | Bachuguda | 6 | 34 | 78,000 | 2,294 | 11.5 | 5,500 | 2 | 0 |
| 6. | Pocharam | 1 | 8 | 5,000 | 625 | 3.0 | 0 | 0 | 0 |
| 7. | Inole | 5 | 47 | 1,00,000 | 2,128 | 15.5 | 5,000 | 3 | 0 |
| 8. | Peddakanjerla | 5 | 39 | 38,000 | 974 | 18.0 | 0 | 0 | 34,800 |
| | Total | 36 | 254 | 4,05,400 | 1,596 | 118.0 | | 8 | |
| *Bollaram Area* | | | | | | | | | |
| 1. | Gandigudem | 5 | 17 | 12,000 | 706 | 8.5 | 7,000 | 9 | 0 |
| 2. | Kistareddypet | 5 | 38 | 60,000 | 1,579 | 5.5 | 0 | 4 | 0 |
| 3. | Sultanpur | 6 | 57 | 24,300 | 426 | 12.5 | 0 | 5 | 0 |
| | Total | 16 | 112 | 96,300 | 860 | 26.5 | | 18 | |
| *Jeedimetla Area* | | | | | | | | | |
| 1. | Kudbullapur | 5 | 36 | 1,30,000 | 3,611 | 21.5 | 0 | 0 | 0 |
| 2. | Ayodhya Nagar | 5 | 20 | 49,800 | 2,490 | 0 | 0 | 0 | 0 |
| | Total | 10 | 56 | 1,79,800 | 3,211 | 21.5 | | 0 | |
| | Grand Total | 62 | 422 | 6,81,500 | 5,667 | 166.0 | | | |

*Source:* Household Survey in the areas of Industrial Estates in Hyderabad, 1995.

Table A 6.4

Incidence of Diseases as Per the Sample Survey of Medak District (1994–5)

| S. No. | Village | Population | Number of Families Surveyed | Water and Air Pollution Diseases | | | | | | | | | | | | |
|---|---|---|---|---|---|---|---|---|---|---|---|---|---|---|---|---|
| | | | | Epilepsy | Asthma | Skin | Throat | TB | Cancer | Conjuncti-vitis | Fever | Perip-heral Neurosis | Respir-atory | Water-borne | Others Un-known | Total Number of Dis-eases |
| 1. | Ganapathigudem | – | 37 | – | 6 | 11 | – | 1 | – | 1 | 6 | 1 | 5 | 2 | 15 | 48 |
| 2. | Pocharam | 1,489 | 45 | 3 | 9 | 8 | – | 1 | – | 5 | 21 | 6 | 6 | 1 | 4 | 64 |
| 3. | Chitkul | 2,670 | 101 | 1 | 16 | – | – | 3 | – | 20 | – | 6 | 9 | 7 | 3 | 65 |
| 4. | Lakdaram | 2,967 | 97 | 1 | 4 | 8 | – | 1 | – | 13 | 3 | 1 | 4 | 4 | 3 | 42 |
| 5. | Ismailkhanpet | 3,851 | 96 | 2 | 2 | 5 | – | 2 | – | 8 | 7 | 4 | 13 | 1 | 6 | 50 |
| 6. | Bachugudem | – | 78 | 1 | 3 | 2 | 2 | – | – | 2 | 3 | 7 | – | – | 2 | 22 |
| | Total | 10,977 | 454 | 8 | 40 | 34 | 2 | 8 | – | 49 | 40 | 25 | 37 | 15 | 33 | 291 |
| *Controlled Villages* | | | | | | | | | | | | | | | | |
| 1. | Ramsanpally | 2,890 | 15 | – | – | 3 | – | 1 | 1 | 2 | – | – | – | – | 2 | 9 |
| 2. | Yarraram | 1,962 | 19 | – | – | 3 | – | – | – | – | – | 2 | – | – | 5 | 10 |
| 3. | Annasagar | 3,820 | 15 | – | – | 2 | – | – | – | – | – | 4 | 2 | – | – | 8 |
| | Total | 8,672 | 49 | – | – | 8 | – | 1 | 1 | 2 | – | 6 | 2 | – | 7 | 27 |

*Source:* District Medical and Health Officer, Medak District.

Table A 6.5
Health Details of the Affected Persons in Three Industrial Estates of Andhra Pradesh

| S. No. Village | Number of Families Surveyed | Members of the Families | Affected Persons | Annual Medical Expenses (Rupees) | Number of Days Hospitalized (Average/ Person) |
|---|---|---|---|---|---|
| *Patancheru Area* | | | | | |
| 1. Lakdaram | 5 | 43 | 8 | 1,200 | 120 |
| 2. Chitkul | 5 | 31 | 6 | 500 | 120 |
| 3. Ganapathigudem | 5 | 28 | 2 | 1,200 | 120 |
| 4. Vadigapalli | 4 | 24 | 0 | 0 | 0 |
| 5. Bachuguda | 6 | 34 | 5 | 2,500 | 84 |
| 6. Pocharam | 1 | 8 | 2 | 1,200 | 120 |
| 7. Inole | 5 | 47 | 11 | 600 | 60 |
| 8. Peddakanjerla | 5 | 39 | 0 | 0 | 0 |
| Total | 36 | 254 | 34 | 7,200 | 624 |
| *Bollaram Area* | | | | | |
| 1. Gandigudem | 5 | 17 | 9 | 1,400 | 120 |
| 2. Kistareddypet | 5 | 38 | 3 | 1,200 | 180 |
| 3. Sultanpur | 6 | 57 | 15 | 1,200 | 120 |
| Total | 16 | 112 | 27 | 3,800 | 420 |
| *Jeedimetla Area* | | | | | |
| 1. Kudbullapur | 5 | 36 | 5 | 1,000 | 90 |
| 2. Ayodhya Nagar | 5 | 20 | 7 | 1,500 | 120 |
| Total | 10 | 56 | 12 | 2,500 | 210 |
| Grand Total | 62 | 422 | 73 | 13,500 | 1,254 |

*Source:* Household Survey in the Areas of Industrial Estates in Hyderabad, 1995.

Table A 6.6

Details of Villages Affected by Pollution and Compensation Paid Towards Damages

| S. No. | Village | Population | 1988–9 | | | 1989–90 | | | 1990–1 | | |
|---|---|---|---|---|---|---|---|---|---|---|---|
| | | | Land Affected | Farmers Affected | Compensation Paid | Land Affected | Farmers Affected | Compensation Paid | Land Affected | Farmers Affected | Compensation Paid |
| 1. | Pocharam | 1,489 | – | – | – | 50.21 | 17 | 51,035 | – | – | – |
| 2. | Bachugudem | 686 | – | – | – | 50.15 | 30 | 43,927 | – | – | – |
| 3. | Lakdaram | 2,967 | – | – | – | 19.34 | 10 | 17,307 | – | – | – |
| 4. | Chitkul | 2,670 | – | – | – | 24.35 | 7 | 21,390 | – | – | – |
| 5. | Peddakanjerla | 1,713 | – | – | – | 108.25 | 44 | 94,702 | – | – | – |
| 6. | Inole | 928 | – | – | – | 24.07 | 14 | 21,081 | – | – | – |
| 7. | Sultanpur | 2,099 | 53.04 | 43 | 19,270 | 42.23 | 58 | 74,982 | 54.14 | 64 | 163,425 |
| 8. | Arutla | 975 | – | – | – | 33.28 | 16 | 29,389 | – | – | – |
| 9. | Byathole | 870 | – | – | – | 144.24 | 40 | 125,666 | – | – | – |
| 10. | Chidruppa | 1,989 | – | – | – | 98.24 | 40 | 84,345 | – | – | – |
| | Total: | 16,386 | 53.04 | 43 | 19,270 | 594.36 | 276 | 563,824 | 54,14 | 64 | 163,425 |

*Source:* District Collector's Office, Sanga Reddy District.

Table A 6.7

Capacity and Cost Details of the Common Effluent Treatment Plants

| Name of the CETP | Year of Commission | Number of Factories in the CETP | Capacity Installed (kl/day) | Capacity Utilized (%) | Capital Cost (Rupees Lakhs) | Operating Costs (Rupees Lakhs/ Month) |
|---|---|---|---|---|---|---|
| JETL, Jeedimetla | 1989 | 80 | 1,050 | 90 | 387 | – |
| PETL, Bollaram | 1993 | 25 | 250 | 64 | 350 | 3 |
| PETLI, Patancheru | 1994 | 46 | 7,500 | 7 | 571 | 2 |
| PTIET, Pallavaram | 1989 | 132 | 30,000 | 45 | 815 | 11 |

*Notes:* JETL: Jeedimetla Effluent Treatment Ltd.; PETL: Progressive Effluent Treatment Ltd.; PETLI: Patancheru Enviro-Tech Ltd.; PTIET: Pallavaram Tanners Industrial Effluent Treatment Co. Ltd.

Table A 6.8

Details About the CETPs Coming up in Various Tanning Centres in Tamil Nadu (As of March 1995)

| S. No. | CETP | Number of Tanneries | Land Area (Acres) | Project Cost (Rupees Crores) | Production Capacity (Metric Tonnes/Day) | Effluent Generated ($m^3$/Day) | Present Position |
|---|---|---|---|---|---|---|---|
| 1. | VANITEC | 96 | 14.0 | 3.37 | 83.6 | 3,120 | Commissioned in Sept. 1991 |
| 2. | AMBURTEC | 44 | 27.0 | 2.98 | 85.0 | – | – |
| 3. | RANITEC | 75 | – | 4.60 | 130.0 | 3,891 | – |
| 4. | PERTEC 1 | 60 | 18.5 | 3.65 | 83.3 | 2,500 | – |
| | PERTEC 2 | 18 | 7.0 | 1.60 | 30.0 | 900 | – |
| 5. | DINTEC | 36 | 50.0 | 4.20 | 80.0 | 2,412 | – |
| 6. | TRYTEC 1 | 7 | – | 1.70 | 35.0 | 1,172 | – |
| | TRYTEC 2 | 10 | – | 1.70 | 34.2 | 1,195 | – |
| | TRYTEC 3 | 13 | – | 1.70 | 33.6 | 1,140 | – |
| 7. | ETET | 77 | 4.0 | 10.00 | – | 8,500 | – |
| 8. | PTIET | 132 | 2.5 | 7.03 | 300.0 | – | Commissioned in Feb. 1995 |

*Source:* Tanners Association, Madras.

*Note:* Entries in column 2 are names of CETPs.

Table A 6.9
Sources of Finance of CETPs in Tamil Nadu (1989–90)

(Rupees Lakhs)

| S. No. | CETP | Members' Contribution | Central Government Subsidy | State Government Subsidy | Loan from Financial Institutions | Total Cost |
|---|---|---|---|---|---|---|
| 1. | VANITEC | 51 | 84.00 | 84.00 | 172 | 391 |
| 2. | AMBURTEC | 60 | 70.00 | 70.00 | 136 | 336 |
| 3. | RANITEC | 92 | 115.00 | 115.00 | 138 | 460 |
| 4. | PERTEC 1 | 73 | 94.00 | 94.00 | 192 | 453 |
| | PERTEC 2 | 32 | 40.00 | 40.00 | 48 | 160 |
| 5. | DINTEC | 30 | 50.00 | 50.00 | 70 | 200 |
| 6. | TRYTEC 1 | 34 | 42.50 | 42.50 | 51 | 170 |
| | TRYTEC 2 | 34 | 42.50 | 42.50 | 51 | 170 |
| | TRYTEC 3 | 34 | 42.50 | 42.50 | 51 | 170 |
| 7. | ETET | 200 | 50.00 | 50.00 | – | 300 |
| 8. | PTIET | 115 | 50.00 | 50.00 | 462 | 677 |

*Source:* Tanners Association, Madras.
*Note:* Entries in column 2 are names of CETPs.

Table A 6.10
Pollutant-specific Charges Levied by CETP at Jeedimetla (1994–5)

| Quality of Industrial Effluent | | Rupees per tanker of 1,000 litres |
|---|---|---|
| COD (ppm) | Upto 5,000 | 175 |
| | 5,001 to 10,000 | 200 |
| | 10,001 to 15,000 | 225 |
| | 15,001 to 20,000 | 250 |
| | 20,001 to 30,000 | 275 |
| | 30,001 to 40,000 | 300 |
| | 40,000 and above | Rs 25 additional charges per tanker for every 1000 units treated. |
| pH | Less than 2 | Not to be accepted |
| | 2 to 3 | 75 |
| | 3 to 5 | 50 |
| | 5 to 6 | 25 |
| | 8 to 9 | 25 |
| | 9 to 11 | 50 |
| | 11 to 14 | 75 |
| | Above 14 | Not to be accepted |
| TDS (Sp.Gr) | Upto 1.05 | Nil |
| | 1.05 to 1.10 | 25 |
| | Above 1.11 | 50 |

*Source:* Contractual Agreement of Industries in Jeedimetla Industrial Estate.
*Note:* ppm: parts per million.

# 7. Collective Action: A Deterrent to Colluding Industry and Corrupt Bureaucracy[1]

M. N. MURTY

Governments of many developing countries and even of some developed countries are malevolent or non-benevolent, so that the responsibility of dealing with market failure and achieving some developmental objectives cannot be left to them. In the case of control of externalities like environmental pollution, the corrupt bureaucracy of developing countries often colludes with the external diseconomy-creating economic agents in augmenting the externality problem rather than controlling it. In such situations, there is no other option left to the parties affected by the externality but to take recourse to collective action or political influence for dealing with the bureaucracy and the perpetrator of the externality. Coase (1960) argues that there is no need for governmental instruments like taxes and subsidies to deal with the externality problem. Given the initial property rights either to the externality-generator or the receivers, the costless bargaining between them in a free market will result in the optimal control of the externality, the final outcome being independent of the initial allocation of property rights. Becker (1983) has shown that the political influence exerted by pressure groups can have similar effects as having a benevolent government to deal with market failure. In his own words:

> The same analysis of competition among pressure groups, without the introduction of social welfare function or benevolent government, explains expenditures on defence and other public goods, taxes on pollution, and other government activities that raise efficiency, even when some groups are hurt by these activities (Becker, 1983).

[1] This chapter is drawn mainly from Murty (1995b).

However, bargaining cannot be costless as Coase originally proposes, and transaction costs are important in the world of Coase and Becker. Even with transaction costs, there can be significant net welfare gains from collective action, and the case for collective action will be reinforced if the government is malevolent. Take the case of industrial water pollution abatement in India. More than twenty years of environmental legislation in India has not produced a pollution tax and the CAC methods used so far by the Indian government have not made a dent in industrial pollution control. This has also been the case with many developing countries. One important reason for this is the absence of awareness among people about the extent of damage from pollution, and their inability to organize themselves for participating in the management of pollution abatement. In such a situation, industry and the non-benevolent bureaucracy have incentives to collude for violating environmental standards. Industry can bribe the regulator to over-report its effluent quality or not report it at all.

There are several agents involved in the political economy of industrial pollution abatement: affected people, elected representatives, bureaucracy and industry. Incentives exist for the emergence of sub-coalitions of agents, like a coalition of affected people and elected representatives and another coalition of bureaucracy and industry. These coalitions can constitute politically-active pressure groups. Murty (1995b) has shown that the competition among these pressure groups may result in the optimal control of pollution.

## 7.1 COASE AND BECKER IN PRACTICE: SOME EXAMPLES FROM INDIA

Empirical findings of a recent survey of some industrial estates in India[2] provides a motivation for the theoretical model provided in this paper. Data from this survey provides insights into the empirical aspects of the economic and non-economic processes leading to collective action in industrial water pollution abatement. It shows that collective action of various agents, like people affected by water pollution, elected representatives, industries, NGOs, and the government is responsible for triggering the processes leading to control

[2] A survey of 17 highly water-polluting industries in India conducted by the Institute of Economic Growth, Delhi in 1996.

of industrial water pollution. It is observed that untreated waste waters from factories in the industrial estates surveyed have been deposited on surface and in local streams, resulting in the degradation of cultivable land and groundwater contamination. Local people have suffered from crop and cattle losses and contracted a variety of water-borne diseases. The government, has failed to take cognizance of these damages, which may support the local peoples' view, that there is a collusion between factories and government officials for sharing the cost saved by factories from non-compliance with pollution standards.

Initial organized efforts by local people through persuasion and even with physical threats to factory owners have not yielded any results. An organized group of local people has had to take recourse to legal action by filling public litigation cases in the courts. There are now a number of successful public litigation cases dealing with industrial water pollution abatement in India. In one such public litigation case (Supreme Court, 1990) concerning an industrial estate in Hyderabad in India, the Supreme Court of India has ordered the constitution of an expert committee (NEERI, 1991) to study the problem of water pollution and to make recommendations about the extent of compensation to be paid to affected people by industries and the remedial measures that industries and the government have to take for preventing water pollution in future. After receiving the committee's recommendations, the Supreme Court has directed factories to pay compensation to the affected people and asked the government to take action as per existing environmental laws against factories so that, they comply with the pollution standards. Political organization of affected people in this case also enlisted the active support of their elected representatives (members of local State Assemblies and the Union Parliament) for the cause of industrial water pollution abatement. Elected representatives in turn have made local pollution control problems part of their election manifestos. There is, therefore, a coalition of affected people and elected representatives to deal with a coalition of factories and bureaucracy for controlling water pollution in this industrial area. Emergence of such an institutional setting, in which there is competition among pressure groups or interest groups, has resulted in the adoption of CETP technology by the factories in each industrial estate. Collective action of affected people has induced factories to organize themselves as a club to construct and manage a CETP.[3] Also, it is interesting to note that now there are many instances in India, in which the government plays a catalytic role rather than its conventional coercive role in environmental management, by providing financial and other incentives to a club of factories for having a CEPT (Misra and Murty, 1995a).

## 7.2 POLLUTION TAXES, CORRUPTION, BRIBES AND PENALTIES

In the case of malevolent government, as it may be in many developing countries, the mere enactment of environmental laws does not guarantee that they are actually executed. More than twenty years of environmental legislations in India (Water Act, 1974 and a number of subsequent legislations) have not helped to have economic instruments like pollution taxes so far and the command and control methods that are normally used by the Indian government could not have any dent in industrial pollution control (Murty, 1995b). This has also been the case with many developing countries. Some of the important reasons for this are the absence of awareness of people about the extent of damage from pollution and their incapability to organize themselves as active political groups to deal with the problem of pollution abatement. In such a situation, industry and bureaucracy have incentives to collude to violate environmental standards. Industry can bribe regulators for over-reporting effluent quality or not reporting at all.

[3] Mr Audun Ruud of Centre for Development and Environment of University of Oslo, while commenting on an earlier version of this paper, has noted that a kind of collective action had actually taken place among people affected by industrial pollution along the Hoogly river in the southern parts of Calcutta before a well-known environmental lawyer Mr M. C. Mehta actually brought these cases to the Supreme Court of India. The court rulings from February 1994 urged several heavily water polluting units along Hoogly river to install required pollution control equipment. Later the West Bengal Pollution Control Board insisted on implementing the court rulings by threatening to close 12 major polluters unless significant steps were taken to reduce water pollution. Thus he feels this is the first instance where collective action succeeded. However, he observes that the political influence of the industrial lobby and the relatively weak position of the environmental lobby significantly influences the speed of the environmental clean-up process.

There can be several points of view about the way corruption takes place in bureaucracy. One viewpoint is that in a pyramidical structure of bureaucracy, corruption increases as one moves down to the lower cadres while it is virtually zero at the highest level. Opportunities for corruption arise when a principal officer delegates enforcement authority to lower-cadre officers (Rose-Ackerman, 1978; Milgrom and Roberts, 1988). In such a system, checks and balances operate to minimize corruption in bureaucracy since there is a probability that the bribe taking lower cadre officer is being caught and penalized by his superior officer. Mukherjee and Png (1994) have shown that in a pyramidical structure of corruption, the penalty for bribe taking and compensation for honest behaviour of lower cadre officers effected by the regulator have uncertain effects on pollution abatement. Another viewpoint is that corruption is distributed (perhaps evenly) among all cadres of bureaucracy. There is tacit agreement among officers of all cadres to share the bribe collected so that a lower cadre officer, say an inspector, has the sanction of his superior to take bribe. It is the bureaucratic corruption of this form that manifests in many developing countries. In India it is a commonly held public belief that inspectors dealing with the collection of taxes, monitoring industrial pollution, etc. have the sanction of their bosses to take bribe. The third viewpoint is that corruption pervades both bureaucracy and politicians (the legislature or elected representatives). There may be a collusion between bureaucracy and elected representatives with the motive of rent seeking for the power constitutionally granted to them. This type of corruption can find a place in societies where the public is not politically active or organized to deal with erring legislators as is the case with democracies in many developing countries. A check on the latter two types of bureaucracy can only come through politically active groups of public or people affected by industrial pollution.

Politically active groups of people affected by an externality can influence their elected representatives in a democracy to stem bureaucratic corruption. They can bring pressure on their elected representatives, say, a Municipal Committee in the case of a local externality, to impose penalties on the bribe taking bureaucracy. Alternatively, they organize themselves as pressure groups to take recourse to legal action against colluding elected representatives, bureaucracy and industry. Thus depending upon the way corruption actually takes place, there can be sub-coalitions of agents relevant for industrial water pollution abatement: a coalition of affected people and elected representatives and a coalition of industry and bureaucracy in the second case; affected people alone and a coalition of industry, bureaucracy and elected representatives in the third case. Collective action of affected people has transaction costs in the form of time devoted and efforts made to organize themselves for detecting violation of environmental standards by a coalition of bureaucracy and industry and for effecting legal action.

## 7.3 PUBLIC PERCEPTION OF DAMAGES AND PENALTIES

Consider a situation in which four sets of agents are involved in the industrial pollution abatement: (a) affected public, (b) elected representatives of public (say a municipal committee), (c) bureaucracy (the executive implementing the environmental legislation) and (d) industry. In the absence of any checks from the affected public, there can be incentives to bureaucracy and industry to violate the environmental standards.[4] Given a pollution tax as per the environmental legislation, industry has an incentive to bribe the bureaucracy, at a rate less than the tax per unit of effluent, for over-reporting its effluent quality. A false certification of higher effluent quality by bureaucracy helps industry to shun its responsibility for meeting the standards. The extent to which the affected public can make bureaucracy and industry comply with environmental standards depends upon its degree of perception of damages and influence it can exert either through political organization or elected representatives, or both. The degree or probability with which the affected people perceive their damage and are able to apprehend the bribe-taking bureaucracy and non-complying industry depends upon the amount of time and effort devoted for politically organizing themselves. Politically active affected people can bring pressure on their elected representatives to impose penalties on bureaucracy and industry if they are caught colluding with each other for violating

[4] The second viewpoint on corruption is chosen here for attempting a detailed analysis. A similar type of analysis can also be made with respect to the third viewpoint on corruption.

the standards. Probability with which they are caught depends upon the degree of political activity.

The following notation is used for developing a model of collective action in the industrial pollution abatement.

$\bar{E}$ : Environmental quality corresponding to statutory level of pollution;[5]

$\hat{E}$ : Actual environmental quality attained by industry;

$E$ : Environmental quality reported by bureaucracy;

$D(E)$ : Actual damages received by affected people, $D'(E) < 0$, $D''(E) > 0$

$A(E)$ : Cost of treatment of effluent to the industry, $A'(E) \geq 0$, $A''(E) \geq 0$

$W$ : Time devoted by affected people for political organization

$V(W)$: Cost of political organization

$\lambda(W)$ : Probability with which affected people (or a fraction thereof) perceive damages from degraded environment: $0 \leq \lambda \leq 1$

$t$ : Pollution tax per unit of effluent

$b$ : Bribe received by bureaucracy from industry per unit of over-reporting of environmental quality

$p$ : Penalty on bureaucracy per unit of under-reporting of environmental quality

$s$ : Penalty on industry per unit of non-disclosure of effluent quality

Now the costs and benefits to various agents involved in the industrial water pollution abatement can be identified as follows:

(a) Perceived damages of affected people

$$\phi(W, E) = \lambda(W)\, D(E) + (1-\lambda(W))\, D(\bar{E}) + V(W) \qquad (7.1)$$

(b) Income to Municipal Committee

$$M(W, E) = \lambda(W)\, (s+p)\, (\hat{E}-E) + t\,(\hat{E}-E) \qquad (7.2)$$

[5] In the case of an optimal pollution tax or Pigouvian tax, the tax liability of a firm is computed as tax rate multiplied by the difference between the actual pollution level and the pollution level with respect to which the marginal damage is zero. However in the case of taxes-and-standards method (Baumol and Oates, 1988), the tax liability of a firm is computed as the tax rate multiplied by the difference between the actual and statutory level (level corresponding to pollution standards) of pollution.

(c) Income to bureaucracy

$$R(W, E) = (b - \lambda(W)p)(\hat{E} - E) \quad (7.3)$$

(d) Cost to factory

$$C(W, E) = A(E) + (b + s\lambda(W))(\hat{E} - E) + t(\bar{E} - \hat{E}) \quad (7.4)$$

Net perceived cost to society can be defined as

$$\pi(W, E) = \phi(W, E) + C(W, E) - R(W, E) - M(W, E)$$
$$= \lambda(W)D(E) + (1 - \lambda(W))D(\bar{E}) + V(W) + A(E) \quad (7.5)$$

Net actual cost to society can be defined as

$$\phi(E, w) = D(E) + V(W) + A(E) \quad (7.6)$$

## 7.4 COASE-BECKER MODEL

Consider a model in which the government has no regulatory role and industrial pollution abatement has to be done through bargaining between affected parties and industry *à la* Coase and Becker. However, there are transaction costs in the form of cost of organization of affected parties V(W) in bargaining with industry which is given to free riding. Fig. 7.1 describes the effects of bargaining

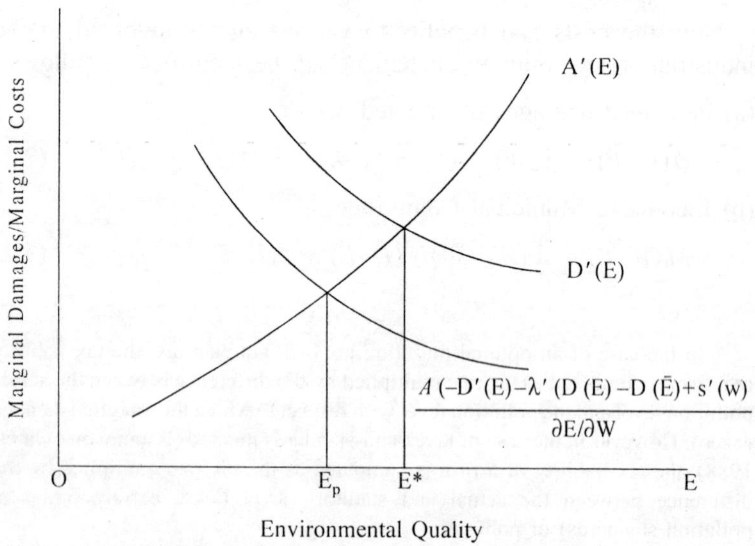


Fig. 7.1 Coase-Becker Model

on the optimal perceived environmental quality. Curves A′(E) and D′(E) represent respectively marginal abatement cost (cost to factory and cost to affected parties) and marginal damage reduction due to improved environmental quality. As political activity or λ increases, the optimal perceived environmental quality increases. Proposition 1: The optimal perceived environmental quality with positive cost of political activity is given by the condition

$$\lambda(-D') = A' + [\lambda' \{D(E) - D(\bar{E})\} + V']/(\delta E/\partial W) \qquad (7.7)$$

and it will be lower than the optimal actual environmental quality. (For proof, see Appendix).

## 7.5 MALEVOLENT GOVERNMENT AND STRATEGIC BEHAVIOUR OF ACTIVE GROUPS IN WATER POLLUTION ABATEMENT

There are incentives for the formation of sub-coalitions of various agents involved in industrial water pollution abatement if the bureaucracy or regulator is corrupt.[6] As explained earlier, it may be possible that affected people and municipal committee act in tandem. People need the help of their representatives to deal with industry and bureaucracy, and the representatives require people's support to get re-elected. Also, it may be possible that industry and bureaucracy have incentives to collude for sharing the costs saved from non-compliance of environmental standards by industry. This simple bifurcation of politics into two sub-coalitions is made to highlight the effect of political activity of pressure groups on industrial water pollution abatement. However, in actuality, a few other sub-coalitions are possible and a similar analysis can be attempted with them. Thus we have two groups of agents:

**Group I** Affected People and Municipal Committee with a strategy $W$ and cost function

$$G^1 = \lambda D(E) + (1-\lambda)\, D(\bar{E}) + V(W) - \lambda(s+p)(\hat{E}-E) - t(\bar{E}-\hat{E}) \qquad (7.8)$$

[6] The structure of corruption in bureaucracy assumed here is different from the one considered in some recent works in this area. It is assumed that the bribes taken by bureaucracy is shared by all cadres of officers, while some recent studies consider a pyramidal structure of corruption (corruption reduces as one climbs up the bureaucratic ladder, becoming virtually zero at the top).

**Group II** Industry and bureaucracy with a strategy $E$ and cost function

$$G^2 = A(E) - \lambda(s+p)(\hat{E}-E) + t(\bar{E}-\hat{E}) \tag{7.9}$$

Given penalties and bribe $(p,s,b)$, Group I minimizes $G^1$ with respect to $W$ given E while Group II minimizes $G^2$ with respect to $E$ given $W$. We have:

$$\frac{\delta G^1}{\delta W} = \lambda'[(s+p)(\hat{E}-E) - (D(E) - D(\bar{E}))] - V'(W) = 0 \tag{7.10}$$

$$(s+p)(\hat{E}-E) - (D(E) - D(\bar{E})) = \frac{V'}{\lambda'}$$

$$\frac{\delta G^2}{\delta E} = A'(E) - \lambda(s+p) = 0 \tag{7.11}$$

$$A'(E) = \lambda(s+p)$$

Equations (7.10) and (7.11) yield the reaction functions of groups I and II which are respectively given by (7.12) and (7.13)[7]

$$W(E),\ W(E) < 0 \tag{7.12}$$

$$E(W),\ E'(W) > 0 \tag{7.13}$$

The equilibrium strategies of two person Nash non-cooperative game are defined as:

$$W^* = W(E^*)$$

and $\quad E^* = E(W^*)$

which are given by the intersection of curves depicting the reaction functions of Groups I and II in Fig. 7.2.

Proposition 2: Increasing penalties $s$ and $p$ on industry and bureaucracy have the effects of increasing environmental quality $E$, given $W$, and decreasing the time devoted to political activity $W$, given the environmental quality $E$. The equilibrium environmental quality of a two person Nash non-cooperative game increases as

[7] In (7.10) $f(W) = V'/\lambda'$ is monotonically increasing in $W$ given $V'' > 0$, $\lambda'' < 0$. Given $\hat{E}$ and $\bar{E}$, $(s+p)(\hat{E}-E) - (D(\bar{E}) - D(E))$ is monotonically decreasing in $E$ provided $(s+p) > D'(E)$. Therefore, we have $W'(E) < 0$. In (7.11) $A'(E)$ and $\lambda$ are respectively increasing in $E$ and $W$. Therefore, given $s$ and $p$, as $W$ increases, $E$ increases $(E'(W) > 0)$.

either at least one of $s$ and $p$ or both $s$ and $p$ increase. (For proof see Appendix).

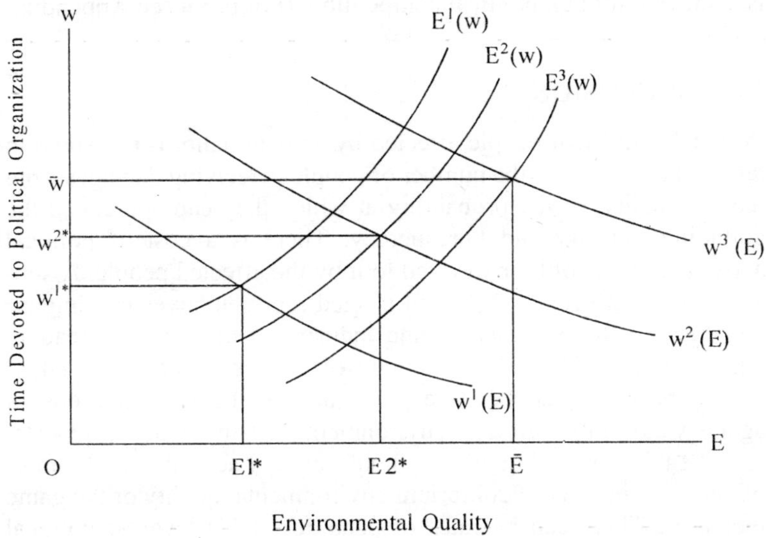


Fig. 7.2 Strategic Behaviour of Active Groups in Water Pollution Abatement

In reality, the capacity of group I to impose penalties on bureaucracy and industry depends upon the degree of political activity within the group. In other words, higher rates of penalties can be feasible with higher $\lambda$.

Increase in either $s$ or $p$, or both, directly increases $E$ given $W$ and indirectly decreases $W$ given $E$. Thus there are welfare gains of increasing penalties either in the form of decrease in the damages due to improvement in environmental quality, or in the form of saving in cost of political activity. Since there are costs to industry from increasing $E$, there can be net welfare gains from increasing penalties so long as incremental damage reductions to affected parties are higher than incremental cost to industry. In fact the optimal rates of penalties and optimal environmental quality are determined by the condition that marginal cost of abatement to industry is equal to marginal reduction in damages. However, in the current framework, the optimal rates of penalties are determined by the statutorily fixed environmental standards.

Proposition 3: Even if Government is malevolent, there exists $W$,s and $p$ with respect to which statutorily fixed environmental quality is achieved through political competition. (For proof see Appendix.)

## 7.6 CONCLUSION

Political activity of people affected by an externality is measured in terms of a fraction or the number of people perceiving damages from the externality, or the probability at which they can apprehend the colluding industry and bureaucracy. There is a cost of political activity in terms of time devoted to it by the affected people. A sub-coalition of affected people and elected representatives can impose penalties on the bureaucracy and industry, if the former is caught taking bribes and the latter is found over-stating the effluent quality.

The Nash non-cooperative game among the sub-coalitions of agents yields an equilibrium environmental quality which is superior to the environmental quality without political activity. As the rates of penalties increase, equilibrium environmental quality of the game increases. There can be rates of penalties and a level of political activity with respect to which statutorily fixed environmental standards can be achieved. Also, the rate of penalty required to achieve the statutory environmental quality through competition between the pressure groups with non-benevolent government is the same as the rate of tax required, using the taxes-standards approach with benevolent government.

# APPENDIX 7

Proposition 1

Proof: Recognizing that $E$ depends upon $W$, we have equation (A.1) from (7.5).

$$\lambda'(D(E) - D(\bar{E})) + (\lambda D' + A')\ \delta E/\delta w + V'(W) = 0 \qquad (1)$$

From (A.1) we have (7.7). Equation (7.7) shows that marginal reduction in perceived damages by affected people is equal to marginal pollution abatement cost to the factory plus marginal cost to affected people. Alternatively, with positive transaction costs and zero liability rule (no liability to factory), marginal cost of abatement to the factory is equal to net marginal damage reduction to affected people. Marginal cost to affected people consists of value of time devoted to political activity and incremental perceived damages due to a rise in political activity (increase in $\lambda$). Fig. 7.1 shows that perceived optimal environmental quality with positive costs of political activity is $E_p$.

If the cost of political activity is zero, we have perceived damages equal to actual damages ($\lambda = 1$, $\lambda' = 0$ and $s' = 0$) implying from (7.7)

$$(-D') = A' \qquad \text{(A.2)}$$

As shown in Fig. 7.1, actual optimal environmental quality, E is higher than perceived optimal environmental quality, $E_p$.

Proposition 2

Proof: From equation (7.12), given $W$, if $s$ and $p$ increase, $E$ should also increase provided $(s+p) > D'(E)$. From equation (7.13) given $E$, if $s$ and $p$ increase, $W$ should fall provided $A'(E) > 0$, $\lambda' > 0$. Equations (7.12) and (7.13) can be alternatively written as

$$\phi_1(E, W) = \frac{V' + \lambda'(D(E) - D(\bar{E}))}{\lambda'(\hat{E} - E)} = s + p \qquad \text{(A.3)}$$

$$\phi_2(E, W) = \frac{A'(E)}{\lambda} = s + p \qquad \text{(A.4)}$$

Given $s$ and $p$, (A.3) and (A.4) will yield reaction curves of Groups I and II as given in Fig. 7.2 above. The equilibrium environmental quality is $E^*_1$ with the reaction curves $E^1(W)$ and $W^1(E)$. If either $s$ or $p$ or both increase, the reaction curve of Group I shifts upwards to $W^2(E)$ and the reaction curve of Group II shifts downward to $E^2(W)$ resulting in the new equilibrium environmental quality $E^*_2$ which is higher than $E^*_1$.

Proposition 3

Proof: If $\lambda$ is equal to 1, Group II has no incentive to overstate the effluent quality. In other words, if $\lambda = 1$, $E = \hat{E}$, there are two cases in which $\lambda = 1$:

(a) costs of political activity are zero, $V(W) = 0$;

(b) costs of political activity are positive and there is high level of political activity or higher $W$ such that $\lambda = 1$.

Also, $\lambda' = 0$ at $\lambda = 1$ since maximum of $\lambda$ with respect to $W$ is 1.

In case (a), we have from (7.10)

$$D(E) - D(\bar{E}) = 0 \tag{A.5}$$

implying $E = \bar{E}$, the statutory environmental quality. We have from (7.11), the rate of penalty $(s + p)$ required to achieve statutory environmental quality as

$$A'(\bar{E}) = s + p \tag{A.6}$$

In case (b), we have from (7.10)

$$D(E) - D(\bar{E}) = V' > 0 \tag{A.7}$$

implying $E < \bar{E}$, or optimal environmental quality with positive costs of political activity is lower than the statutorily fixed environmental quality. Given that $V$ can be negligibly small in relation to $D(E)$, the difference between $E$ and $\bar{E}$ can be negligible.[1] From (7.11) we have (A.6) again.

Note also that if the government is benevolent, the taxes-standards approach requires the pollution tax $(t)$ to be fixed such that $A'(\bar{E}) = t$. In other words, the rate of penalty required to achieve the statutory environmental quality through competition between pressure groups with a non-benevolent government is the same as the rate of tax required using the taxes-standards approach with a benevolent government.

[1] Suppose $V(W) = aW$, where 'a' is per unit value of time or wage for per person day, we have $\surd = a$ and $D(E) - D(\bar{E}) = a$. Given that 'a' is negligibly small in relation to $D(E)$, there will be a negligibly small difference between $E$ and $\bar{E}$ so that

$$D(E) - D(\bar{E}) = a$$

# PART III
# Accounting for Environmental Costs and Benefits

# Introduction to Part III

Valuation of natural resources in general and environmental resources in particular poses formidable problems because of missing markets. Valuation of these resources is needed for two important reasons:

- to justify the investments or defensive expenditures made in relation to benefits from preservation.
- to develop a generalized system of economic and natural resource accounting for estimating the environmentally corrected GDP.

For facilitating the choice between developmental uses and preservation uses of environmental resources, estimates of benefits and costs of alternative uses are needed. Development benefits like power, irrigation, fisheries etc. are private goods which can be estimated from the market data. On the other hand, preservation benefits like amenity services, biodiversity, etc. are non-marketed goods. The preservation benefits are variously classified as user benefits, option value benefits (Weisbrod, 1964) and existence and bequest value benefits (Krutilla, 1967). The choice between developmental and preservation alternatives requires the evaluation of benefits from both, since one becomes the opportunity cost of the other (Krutilla and Fisher, 1975). Preservation benefits are positive externalities of environmental resources, and as mentioned above, the markets for them are absent. And without a market, there are no price and quantity data to estimate these benefits. There are methods specially designed to estimate the preservation benefits which are classified as methods based on observed behaviour and hypothetical methods. The methods based on observed behaviour include travel cost, hedonic property values, etc. while the hypothetical methods include contingent valuation and various variants of it. While the travel cost and hedonic prices methods can be used to measure only direct user benefits like recreational benefits, the contingent valuation methods can be used to measure both user and non-user (existence and bequest values) benefits. Contingent valuation methods are used not only for valuing environmental externalities like recreation, biodiversity etc., but also for valuing private goods like health, water supply, education, etc. for which markets may be imperfect or incomplete.

Take for example the measurement of economic benefits that an individual receives from the improved water quality. These can be measured either as cost savings (savings in labour time, energy etc.) or as the consumer surplus associated with better water quality. It can be argued that the benefits an individual receives depends on his preferences measured in terms of money that he would be willing to pay (WTP) in order to receive the services of better water quality. There are broadly three ways of estimating what people are willing to pay for better water quality, (a) by observing prices that people pay for goods in various markets; (b) by observing individuals' expenditures which may be complementary to water quality (for example, swimming, sports fishing etc.) or time or labour devoted to avoid the loss from inferior water quality and (c) by asking people directly what they are willing to pay for the improved water quality. While the first two approaches are based on observations of behaviour and revealed preferences, the third technique is based on stated preferences using the contingent valuation method.

The contingent valuation method (CVM) has distinct advantages over the methods based on observed behaviour. CVM can be used to value services that are normally very difficult to assess with indirect approaches and it can elicit information on willingness to pay for different levels of water quality according to a range of options offered. However, there is a lack of familiarity with CVM surveys in developing countries. Also CVM surveys are neither cheap nor easy to conduct and CVM results from one location cannot be applied to another. Further, it is important to control strategic, compliance, hypothetical and sample-size biases that arise in such surveys (Mitchell and Carson, 1989).

Valuation of preservation benefits of environmental resources provides us with an estimate of opportunity cost of unsustainable development (development without preservation). Alternatively, a strategy for sustainable development requires an estimate of benefits from the preserved environmental resources. Sustainable development which guarantees the sustainability of combined ecological and economic systems is the macro goal of ecological economics, a transdisciplinary science addressing the relationship between ecosystems and economic systems (Costanza, 1996). There are two viewpoints about sustainability: weak sustainability and strong sustainability. Weak sustainability presumes that technological progress ultimately removes all resource constraints to economic growth and man-made capital can be substituted by natural capital. On the other hand, the strong sustainability maintains that man-made capital cannot be substituted by natural capital and there are limits imposed by natural resources on development. Ecological economics defines sustainability as the amount of consumption that can be continued indefinitely without degrading stocks of man-made and natural capital. This definition draws on definitions of sustainability in ecology and economics. In ecology, sustainability reflects prudent behaviour by a predator that avoids overexploiting its prey to ensure an optimum sustained yield (Odum, 1971). In economics, income (national income) is defined as the maximum amount a person (population) can consume during a period of time without impoverishing himself (itself) (Hicks, 1946). The definition of sustainable income in ecological economics requires the maintenance of man-made capital and natural capital at certain levels.

There are two methods of measuring sustainable income: net price method and user cost method (May and Motta, 1996). The net price of a depleted resource is its present value which is equal to revenue or rent accruing to the owner of the resource. Assuming that this rent reflects the depletion of the resource, it has to be deducted from gross income in the same way as depreciation is deducted from value-added in manufacturing activities. Reppeto et al. (1989) and Solorzano et al. (1992) followed this procedure in evaluating the asset loss derived from oil and timber extraction respectively in Indonesia and Costa Rica. User cost method assumes that man-made capital is substitutable for natural capital and uses the definition of true income, that is, the level of consumption that will not harm future use. A finite series of anticipated revenues from the resources being depleted must be converted into an infinite series of true income in such a way that the current values of both are equal. That is, part of the anticipated revenue would be set aside regularly as an investment capable of generating a perpetual income stream equivalent to true income. This revenue set aside is called user cost of the depleted resource which has to be deducted from GDP. These two methods take two extreme views on the valuation of natural resources, the former considering the entire rent from the resource as depreciation of natural capital and the latter assuming perfect substitutability between man-made and natural capital. These methods are appropriate for the valuation of exhaustible resources but not for renewable environmental resources like water, air and forests.

The System for Integrating Environmental and Economic Accounting (SEEA) of United Nations takes the view that man-made capital and natural capital are not substitutes, since certain types of services like recreation, biodiversity, etc. are unique to natural resources. It prescribes that the maintenance costs and damage costs from depletion of environmental resources have to be deducted from GDP. The main objectives of SEEA are given as follows: (i) segregation and elaboration of all environmental related flows and stocks of traditional accounts; (ii) linkage of physical resource accounts with monetary environmental accounts and balance sheets; (iii) assessment of environmental costs and benefits; (iv) accounting for the maintenance of tangible wealth and (v) elaboration and measurement of indicators of environmentally adjusted product and income. The SEEA recommends three categories of monetary valuation of environmental resources: One version applies a market valuation approach which rearranges only environmental changes already contained in the asset accounts of the conventional system of national accounts. The second version uses maintenance valuation approach which estimates the costs that would have been required to keep the natural environment intact during the accounting period. The third version combines the market valuation with the contingent valuation approach. The second and the third valuation approaches recognize that the costs required to keep natural resources intact have to be accounted for and benefits from environmental services for which markets are absent have to be measured using contingent valuation methods.

# 8. Environmentally Adjusted GDP: Theory and Illustrations from Indian Industry[1]

M. N. MURTY

Resource-accounting problems and their solutions become more transparent if a distinction is made between environmental resources and exhaustible resources. Environmental resources are renewable resources, or resources which are regenerative upto certain levels of use, beyond which irreversibility may set in, making them exhaustible. Services provided by environmental resources can be classified as services for which markets are present, and services for which markets are absent. In other words, they supply both private and public good type of services. On the other hand, exhaustible resources like fossil fuels, minerals and metals are private goods, or goods for which markets are present. The conventional system of national accounts (CSNA) accounts for all goods and services for which markets are present. It is only goods and services for which markets are absent that pose formidable problems in resource accounting. Environmental resources supply amenity services, biodiversity, soil conservation, waste disposal services, etc. for which markets are absent. Recent studies on resource accounting (Dasgupta and Maler, 1990; Ministry of Finance, Stockholm, 1991; UN, 1993; Dasgupta, Kristrom and Maler, 1994) have provided good insights into the problem of integrating natural resource accounting with a system of national accounts. Even though some recent empirical studies (Repetto et al., 1989; Bartelmus et al., 1992; Vincent, 1993; Van Tongeren et al., 1991) provide some objective methods of resource accounting, the methods used for valuing public goods type of environmental services are far from satisfactory.

[1] This chapter is drawn mainly from Murty (1996b).

Current literature on Environmental Economics provides methods like contingent valuation, travel cost, hedonic prices, etc. for valuing environmental services and a large number of micro-level case studies using these methods are now available. Integration of environmental resource accounting with a system of national accounts requires information, generated through a large number of micro-level studies dealing with water resources, atmosphere and forests in diverse climatic conditions and geographical locations. Such a comprehensive set of studies were not yet made even in the developed countries, not to speak of developing countries.

Estimates of GNP and its rate of growth based on CSNA can be higher than their environmentally corrected counterparts. In some countries, the rate of growth of per capita environmentally corrected net national product (ENP) can be negative and environmentally corrected gross national product (EGNP) can be nil (Dasgupta and Maler, 1990). The differences between GNP and EGNP depends upon, among other things, the following:

• current state of the economy in terms of preservation and pollution abatement activities and current levels of degradation (resources degraded upto a point of irreversibility, resources degraded but retrievable by abatement activities and resources which are not degraded at all)

• the extent to which the services of environmental resources with missing markets can be taken into account in CSNA.

In an economy having certain pollution abatement and preservation activities, there are benefits from private and public goods supplied by undegraded environmental resources, restored environmental resources through abatement activities, and costs in terms of cost of abatement and preservation, and damage from resources degraded upto the point of irreversibility. In the measurement of GNP using CSNA, account is taken of the value-added benefits from the private goods supplied by undisturbed and preserved environmental resources and the value-added losses from the fall of private goods supply from the degraded resources, while the value-added benefits from the supply of public goods including value-added losses from the reduced supply of public goods attributed to degraded resources are not taken into account. In the measurement of EGNP, the value-added benefits from the net supply of public goods from the environmental resources are accounted. In this case, EGNP can be obtained by adding to GNP the value added through the net supply of public goods from environmental resources. The net supply of public goods can be positive or negative depending upon the state of the economy. In an economy in which there are no preservation and abatement activities and the resources are degraded upto the point of irreversibility, the net supply of public goods can well be negative.

Maximization of EGNP can be an objective in making the choice between the developmental and preservation uses of environmental resources. Consider an economy with and without preservation of environmental resources. The opportunity cost of preservation is the value-added foregone from the developmental use of environmental resources. The EGNP with preservation is higher than, equal to or lower than EGNP without preservation according to whether the benefits from preservation are higher than, equal to or lower than the benefits from development uses. Preservation has to be carried upto the point at which the incremental EGNP (difference between marginal benefits and marginal costs of preservation) is zero.

## 8.1 SUSTAINABLE SOCIAL PRODUCT: ACCOUNTING PRICES FOR ENVIRONMENTAL RESOURCES

Ecological economics defines sustainable social product as the amount of consumption that can be continued indefinitely without degrading the capital stock, including the natural capital stock. There are two concepts of sustainable development: weak sustainability and strong sustainability. The concept of weak sustainability assumes that man-made capital is substitutable for natural capital, and that technological progress makes this feasible, while the concept of strong substitutability maintains that man-made capital is not substitutable for natural capital, and that there are limits imposed by natural resources on development. Ecological economics recognizes that man-made and natural capital are not substitutable beyond certain levels, and in fact, they are complementary to each other in promoting sustainable development.

Social product or environmentally corrected national product can be measured using the income approach in national income accounting by considering environmental resources as primary inputs along with other primary inputs like labour and capital. EGNP can be estimated, given a set of shadow prices for all primary inputs including environmental resources. The shadow prices of labour and capital can be defined in terms of value-added attributable to them in national income accounting. It is difficult to estimate the shadow prices of environmental resources since (a) they offer public goods type of services, and (b) they often affect EGNP directly as a stock, and not merely as service flows.

The definition of stock may differ for different resources. In the case of water resources, the stock may be defined in terms of both quantity and quality (quality is inversely related to the levels of BOD, COD, SS, etc. beyond the threshold or natural absorptive levels) of water while in the case of atmosphere, the stock may be in terms of quality (quality is inversely related to levels of $CO_2$, $SO_2$, $NO_2$, SS, etc. beyond threshold levels) only. In the case of forest resources, the stock may be expressed in terms of area under forest cover. Fisheries and aquifers are useful not only for the harvest they provide (this is the flow), but also as a stock, because harvesting and extraction costs are low if stocks are large. Tropical forests are beneficial not only for the timber they supply (this is the flow service), but also as a stock since they prevent soil erosion and, in the case of large tropical forests, help maintain a varied genetic pool and contribute substantially to the recycling of carbon dioxide (Dasgupta and Maler, 1990). Similarly in the case of a water body (surface water or aquifers), it not only provides water for household and industrial purposes (this is a flow), but the quality of raw water as a stock affects the cost of water supply. The same problem arises with atmospheric quality as a stock; the presence of $CO_2$, $SO_2$, $NO_2$ and SSP in the atmosphere will cause loss of health and climatic changes. Climatic changes may result in loss of productive land and crops due to flooding. Dasgupta, Kristrom and Maler (1994) have provided a method of defining the shadow prices of environmental resources taking into account their aforementioned characteristics. In particular, they have shown that ENP can be measured given the shadow prices of primary resources, including environmental resources.

Environmental resources enter as inputs in the production functions for the activities supplying private and public good type of services. Therefore, shadow prices of environmental resources may be defined as the value-added attributable to them in the activities in which they figure as primary inputs. The value added through the supply of private goods like water for industrial and household use, fish production, etc. can be estimated using market prices. The value added through the supply of public goods, like civic amenities, waste-disposal services, etc. has to be estimated using prices obtained through valuation methods which use artificially created markets.

Valuation of quality of environmental resources can be made in terms of marginal gains due to improvement of environmental quality, say reduction in BOD or COD by one unit in a water body or, reduction of $CO_2$, or $SO_2$ by one unit in the atmosphere. The marginal gains from the improved environmental quality can be measured in terms of the sum of the increase in value-added in various sectors in the economy due to one unit increase in the quality of environment.

The ENP can be measured as

$$ENP = w.L + r.K + p_1.E_1 + p_2.E_2 + p_3.E_3 \quad (8.1)$$

where $L, K, E_1, E_2, E_3$ are primary inputs like labour, capital, water resources, forests and atmosphere, with $w, r, p_1, p_2$ and $p_3$ being their respective shadow prices.

The effect of preservation of environmental resources on ENP is positive only if the benefits from preservation exceed the costs of preservation. The choice between the preservation and development uses of an environmental resource has to be made depending upon its effect on ENP. Consider two hypothetical situations of an economy with preservation and without preservation. Let the shadow prices of primary inputs denoted above be distinguished by superscript '0' for a situation without preservation and superscript '1' for a situation with preservation. Thus we have $w^j$, $r^j$, $p^j{}_1$, $p^j{}_2$ and $p^j{}_3$, $j = 0, 1$. The ENP is given as

$$ENP^j = w^j.L + r^j.K + p^j{}_1.E_1 + p^j{}_2.E_2 + p^j{}_3.E_3,\ j = 0, 1 \quad (8.2)$$

Assuming that the value-added arising out of developmental use of environmental resources is attributed only to labour and capital and there are no values placed on environmental resources in a situation without preservation, we have

$$w^0 > w^1,\ r^0 > r^1,\ p^{0i} < p^{1i},\ i = 1,2,3. \quad (8.3)$$

$ENP^1 \gtreqless ENP^0$ according to

$$[(w^1 - w^0)\,L + (r^1 - r^0)\,K] + \sum_{i=1}^{n} [(p_i^1 - p_i^0)\ E_i] \gtreqless 0 \quad (8.4)$$

given L, K, and $E_i$, $i = 1, 2, 3$.

In an economy with preservation, part of $L$ and $K$ are used for preservation which in turn contribute to increased productivity of environmental resources ($p_i^0 < p_i^1$, $i = 1,2,3$). The reduction of value-added in the private good production as a result of preservation is the opportunity cost of preservation. The fall in value-added in private good production is reflected in lower prices of labour and capital with preservation ($w^0 > w^1$, $r^0 > r^1$). Therefore, the first term in square brackets in (4) represents the cost of preservation which has to be weighed against the benefits from preservation, the second term in square brackets.

ENP can also be estimated using the expenditure method for national income accounting. Thus

ENP = consumption of marketed goods + public expenditures
+ defensive expenditures against damages to the flow of environmental amenities
+ the value of change in the capital stock including the stock of environmental defensive capital
+ value of net flow of environmental amenities and other services of consumptive type
+ value of net change in the environmental resource base

Dasgupta and Maler (1994, 1997) have also considered a similar method for calculating ENP, except that they recommend a different treatment of the wage bill; the deduction of the wage bill of raw labour (assuming that wages are equal to marginal ill-being of work effort) and addition of investment in human capital development (say opportunity cost of time for students) in the calculations.

Benefits from environmental amenities and biodiversity are in the form of direct user benefit, option value, existence value and bequest value[2]. Simulated or artificial markets for these services have to be used for valuing them, and the contingent valuation method uses such markets. Travel cost and hedonic prices methods use surrogate markets for environmental amenities for measuring direct user benefits. The measurement of value-added by environmental resources in waste-disposal activities also poses some difficult problem. To understand these problems, let us consider a water resource, a lake, and consider two distinct situations (Murty, 1995a). The first assumes that the lake is used for waste-disposal purposes only (the factory draws water from other sources for use in its production

[2] Refer ch. 9.

processes). In the second situation, there is no lake. Where the lake exists, the value it adds is the saving to the factory in waste-disposal costs. Since the lake is considered a stock like capital, the savings in pollution abatement costs to the factory may be treated as the value or cost of capital services for a stock of capital. However, unlike capital, an environmental resource, like the lake, is not owned by the factory, but belongs to the society. The pertinent question is therefore whether the value added by the lake can be treated as part of the private profits of the factory or whether the government ought to collect from the factory a royalty for the waste-disposal services rendered by the lake. Further, if the ambient water quality standards of the lake fall short of the prescribed pollution standards, so that the factory has to incur certain abatement costs. The value added by the lake amounts to the difference between the current abatement costs and the cost of the next best alternative abatement method. In this case, any payment made by the factory will consist of a royalty plus a pollution tax or cost of abatement. In a free market economy however, any factory can use a given water body to dispose waste without paying for it. Only the imposition of fiscal instruments such as pollution taxes or any other form of institutional arrangements, will force the factory to pay for using the lake for waste-disposal.

## 8.2 A METHOD OF MEASURING VALUE-ADDED BY ENVIRONMENTAL RESOURCES

The production approach in national income accounting may be a convenient method for estimating ENA. The method can be better understood by considering two distinct states of the economy. The first is the state of the economy in which activities facilitating preservation and pollution abatement are present and there is no resource degradation and the second is the state of the economy in which these activities are absent.

In CSNA, the value added by a production activity $i$, say $VA^i$, is measured as

$$VA^i = Y^i - M^i, \; i = 1\ 2, \ldots, N \qquad (8.5)$$

where

$Y^i$: Value of output of $i$th activity
$M^i$: Intermediary input cost of $i$th activity

and the total value added (VA) is measured as

$$VA = \sum_{i=1}^{n} VA^i \tag{8.6}$$

In an economy in which there are preservation and pollution abatement activities, CSNA accounts for the cost of abatement and preservation, as well as the benefits from the services of environmental resources for which markets are present. For instance, the benefits of supplying water for household and industrial use, fish production and navigation from a preserved water body are taken into account in the calculation of value-added by the production activities supplying these services. The costs of abatement and preservation are also taken into account in the calculation of value-added by water polluting industries. In this case, the value added by *the i*th activity is measured as

$$VA_2^i = Y^i - M_1^i - M_2^i \tag{8.7}$$

where

$M_1^i$: intermediary input cost of main plant of *i*th activity, and
$M_2^i$: intermediary input cost of pollution abatement of *i*th activity.

In an economy having preservation and pollution abatement activities, CSNA underestimates GDP or value-added in the economy. First of all CSNA does not consider environmental resources as primary inputs. The entire value-added in the environmental services for which markets are present is attributed to conventional primary inputs, labour and capital. Secondly, it does not account for the environmental services for which markets are absent.

Environmental services like amenities, biodiversity and waste-disposal are public goods. The measurement of value-added benefits from these services requires the definition of the control region of a resource. For example, the control region of a water resource is the area covering all the activities using that resource for waste-disposal. Therefore, pollution abatement and preservation attempts by all these activities result in the preservation of the water resource which supplies certain public goods. In order to provide a method for estimating environmentally-corrected value-added (EVA) consider an economy having *S* number of water resources, each having a control region as defined above. Assume that the *s*th water resource has *N* number of activities using it for waste-disposal in its control region. The value added by the activities in the *s*th control region

($EVA_s$), is measured by

$$EVA_s = \sum_{i=1}^{Ns} Y_s^i - \sum_{i=1}^{Ns} M_{s1}{}^i + \left(Z_s - \sum_{i=1}^{Ns} M_{s2}{}^i\right); \; s = 1, 2 \ldots s \quad (8.8)$$

where

$Y^i{}_s$: Value of main product of $i$th activity in $s$th region.

$Z_s$: Value of public goods type of services by $s$th water body

$M^i{}_{s1}$: Value of intermediary inputs in the main plant of $i$th activity in $s$th region

$M^i{}_{s2}$: Value of intermediary inputs in the preservation or pollution abatement in $i$th activity of $s$th region.

The environmentally-corrected value added is given by

$$EVA = \sum_{i=1}^{S} EVA_s \quad (8.9)$$

Alternatively, EVA can be defined as

$$EVA = \sum_{i=1}^{N} VA_2^i + \sum_{s=1}^{S} Z_s$$

$$where\ N = \sum_{s=1}^{S} N_s$$

Therefore in the state of an economy having preservation and pollution abatement activities, EVA estimated using CSNA is lower than EVA.

In a state of the economy in which preservation and pollution control activities are absent, CSNA takes account of losses in environmental services for which markets exist, but fails to account for losses in public good type of environmental services. The environmentally adjusted value added (EVA) in this case is given by

$$EVA = \sum_{i=1}^{N} Y^i - \sum_{i=1}^{n} M_1^i \sum_{s=1}^{S} Z_s \quad (8.10)$$

Therefore in this case, VA estimated using CSNA is higher than EVA.

## 8.3 UNITED NATIONS' SYSTEM OF INTEGRATED ENVIRONMENTAL AND ECONOMIC ACCOUNTING

The United Nations' System of Environmental and Economic Accounting (SEEA) deals with both the concept and measurement of sustainable economic growth, and the valuation of environmental resources. The main objectives of SEEA are given as (i) segregation and elaboration of all environment-related flows and stocks of traditional accounts; (ii) linkage of physical resource accounts with monetary environmental accounts and balance sheets; (iii) assessment of environmental costs and benefits; (iv) accounting for maintenance of tangible wealth and (v) elaboration and measurement of indicators of environmentally-adjusted product and income.

The SEEA uses three categories of monetary valuation of environmental resources. One version applies a market-valuation approach which rearranges environmental changes already contained in the asset accounts of the conventional system of national accounts. The second version uses maintenance valuation approach, which estimates the cost that should have been required to keep the natural environment intact during the accounting period. The third version combines market valuation with contingent valuation approach. SEEA is developed on the premise that man-made capital and natural capital are not substitutable, since certain types of services like recreation, biodiversity, etc. are unique to natural capital. The second and third valuation approaches prescribed by SEEA recognize that the costs that are incurred to keep natural resources intact have to be taken into account, and benefits from environmental services for which markets are absent have to be measured using contingent valuation and other methods. The following section provides the estimates of costs that would be required for Indian industry to keep the quality of water resources intact, satisfying the prescribed standards using the maintenance valuation approach of SEEA.

## 8.4 COST OF WATER POLLUTION ABATEMENT IN INDIAN INDUSTRY

Legislations and development of institutions to deal with environmental pollution in India have led many industries to take recourse to pollution abatement (Central Pollution Control Board, 1989; Murty, 1995a). An overview of the Indian economy today reveals that there are pollution abatement and preservation activities in industry, agriculture and forestry which should not be overlooked in the measurement of GNP. A recent survey[3] shows that many

[3] Murty et al., 1995, A survey of 17 highly water polluting industries in India.

factories belonging to the 17 highly water polluting categories of industries have made significant investments in pollution abatement (Table 8.1). Current methods of national income accounting practised in India do not completely account for the benefits and costs of such investments. As already pointed out in the foregoing section, while benefits from the supply of private goods from the preserved resources are taken into account, benefits from the supply of public goods are not. Environmental resources are not treated as primary inputs. The entire value-added in a water polluting industry is attributed to conventional primary inputs, capital and labour. Rate of return on capital is computed assuming that the entire capital stock of the factory including the capital stock of the effluent treatment plant (ETP) is used to produce conventional output only.

Table 8.1
Capital Costs and Recurring Expenditures for Pollution Abatement in Some Major Industries in India (Rs Million)

| Industry | Capital Costs Incurred so far | Recurring Costs |
|---|---|---|
| 1. Basic Industrial Chemicals | 7.5 | 1.00 |
| 2. Man-made Fibre | 57.8 | 11.80 |
| 3. Cement | 5.3 | 2.44 |
| 4. Sugar | 7.0 | 1.08 |
| 5. Batteries | 11.0 | 0.33 |
| 6. Fertilizers and Pesticides | 12.1 | 4.48 |
| 7. Paper | 50.7 | 13.93 |
| 8. Cement | 61.1 | 0.06 |
| 9. Chemicals | 36.6 | 2.00 |
| 10. Pulp and Paper | 39.0 | 3.50 |

*Source:* FICCI Survey results reported in Department of Scientific and Industrial Research, Government of India, 1993.

In an economy with full employment, there is a trade-off between the production of conventional output (marketable private goods like chemicals, processed leather, sugar, etc.) and pollution abatement and preservation. For example, a factory has a choice of using scarce labour and capital either for producing more of its conventional output, or for having an effluent treatment plant (ETP). The opportunity cost of pollution abatement in a factory may be expressed as value-added losses as a result of using capital and labour in the ETP instead of in the main plant for producing conventional output. It may be useful to estimate this opportunity cost for Indian industry. An attempt is made in the following pages to estimate the opportunity cost of water pollution abatement for the Indian industry in hypothetical situation in which all the industries practise pollution abatement on the lines of current pollution abatement practices of a representative factory belonging to each industry. Sources of data used for this analysis are the recent survey of 17 highly water polluting industries in India and the Annual Survey of Industries (ASI).

The current status of effluent treatment plants for water pollution abatement in some major industries in India is such that not even 25 per cent of factories are meeting the pollution standards. For example in the case of the sugar industry, out of a total of 363 units, only 65 units are meeting the standards, 110 factories are undertaking only partial treatment and 73 factories are still constructing effluent treatment plants (Government of India, 1993). A survey by FICCI of 400 companies which have been undertaking effluent treatment, reveals that on an average these companies have spent Rs 27 million each on capital stock and incur a recurring cost amounting to Rs 4.5 million on pollution abatement.

Table 8.1 provides information about capital costs so far incurred and recurring expenditures made for pollution abatement by some major industries in India.

Data from a recent survey (Murty et. al., 1995) of 100 factories belonging to the category of 17 highly water polluting industries in India are used to estimate the capital equipment, primary input cost, cost of material and power, as percentage of turnover for each polluting industry as reported in Table 8.2 below.

Using these estimates and ASI data for the turnover of each industry for the financial year 1991–2, estimates of capital stock, primary input cost, material and power cost, and total cost for water pollution abatement for each highly water-polluting industry in India are obtained (see Table 8.3).

The estimates in Table 8.3 are obtained for a hypothetical situation in the Indian economy in which all the factories in an industry invest in water pollution abatement. The estimated capital stock for water pollution abatement in the Indian industry is Rs 59,179.18 million. The primary input cost (cost of labour and capital) and total cost (primary input cost plus the cost of material and power inputs) of water pollution abatement for Indian industry are estimated at

Rs 17,748.66 million and Rs 21,210.36 million respectively. These estimates are made at 1993–4 prices for the levels of industrial production given by ASI data for 1991–2.

Table 8.2
Estimates of Capital Stock, Value-added Foregone, Material and Power Cost, and Total Annual Cost for Water Pollution Abatement as Percentage of Turnover in Major Water-polluting Industries in India

| Industry | Capital Stock | Value-added Foregone | Material & Power Cost | Total Abatement Cost |
|---|---|---|---|---|
| 1. Fertilizer | 3.03 | 0.53 | 0.12 | 0.65 |
| 2. Sugar | 0.77 | 0.09 | 0.10 | 0.19 |
| 3. Cement | 0.03 | 0.01 | 0.00 | 0.02 |
| 4. Distillery | 4.78 | 2.04 | 0.68 | 2.72 |
| 5. Chemicals | 21.50 | 6.72 | 0.18 | 6.90 |
| 6. Thermal Power | 0.52 | 0.08 | 0.69 | 0.77 |
| 7. Refinery & Petroleum products | 1.12 | 0.35 | 0.06 | 0.41 |
| 8. Tannery | 2.68 | 0.87 | 0.13 | 1.01 |
| 9. Zinc | 0.87 | 0.15 | 0.14 | 0.30 |
| 10. Iron & Steel | 0.22 | 0.05 | 0.06 | 0.11 |
| 11. Paper | 6.78 | 1.61 | 0.76 | 2.37 |
| 12. Drugs | 5.85 | 2.20 | 0.65 | 2.85 |

*Source:* Estimated as explained in the text.

Primary input cost of Rs 17,748.66 million for water pollution abatement in the Indian industry may be interpreted as value-added foregone due to water pollution abatement. In other words, this is the income that should have been generated if capital and labour used in effluent treatment are alternatively used in the production of conventional output. In reality, only one-third or one-fourth of factories in Indian industry are practising water pollution abatement, the cost (especially the intermediary input cost) of which is partially taken into account in the current estimates of national income using CSNA. However, CSNA does not recognize that the same conventional output currently produced by the industry may also be produced without effluent treatment by these factories if there were no environmental concern in India. Therefore, the value-added loss due to water pollution control currently practised by the Indian industry is only one-third or one-fourth of value-added loss of Rs 17,748.66 million estimated for a hypothetical situation.

Table 8.3

Estimates of Capital Stock, Value added Foregone and Total Annual Cost for Water Pollution Abatement in Major Water Polluting Industries in India

| Industry | Turnover | Capital Stock | Wage Bill | Material Cost | Power Cost | Capital Cost | Value-Added | Total ETP Cost |
|---|---|---|---|---|---|---|---|---|
| 1. Fertilizer | 120,426.70 | 3,653.59 | 19.56 | 103.89 | 37.86 | 621.11 | 640.67 | 782.42 |
| 2. Sugar | 85,120.30 | 654.08 | 28.19 | 20.38 | 61.29 | 52.33 | 80.52 | 162.19 |
| 3. Cement | 64,477.30 | 20.40 | 0.91 | 0.00 | 1.36 | 7.55 | 8.46 | 9.82 |
| 4. Distillery | 16,627.90 | 794.23 | 21.91 | 61.08 | 51.29 | 317.69 | 339.60 | 451.97 |
| 5. Chemical | 199,208.80 | 42,828.92 | 111.18 | 148.60 | 200.29 | 13,276.97 | 13,388.15 | 13,737.04 |
| 6. Thermal Power | 243,764.70 | 1,279.18 | 60.10 | 1,502.56 | 190.32 | 127.92 | 188.02 | 1,880.90 |
| 7. Refinery & Petroleum Products | 126,695.40 | 1,424.25 | 21.72 | 48.32 | 26.88 | 427.28 | 449.00 | 524.20 |
| 8. Tannery | 14,874.60 | 399.11 | 18.27 | 9.08 | 10.58 | 111.75 | 130.02 | 149.68 |
| 9. Zinc | 7,506.70 | 65.30 | 1.57 | 10.44 | 0.44 | 9.80 | 11.37 | 22.25 |
| 10. Iron & Steel | 200,119.70 | 443.66 | 54.24 | 32.54 | 94.91 | 39.93 | 94.17 | 221.62 |
| 11. Paper | 47,312.90 | 3,209.73 | 54.58 | 117.07 | 244.28 | 706.14 | 760.72 | 1,122.07 |
| 12. Drugs | 75,337.60 | 4,406.73 | 115.60 | 262.48 | 225.76 | 1,542.36 | 1,657.96 | 2,146.20 |

*Source:* Estimated as explained in the text.

In the generalized system of national accounts as described in Section 3, the cost of preservation or pollution abatement is attributed to the supply of environmental services which also contributes to value-added in the economy. In a full employment economy, there is a choice of using capital and labour between the supply of conventional output and the supply of environmental services at the margin which has to be guided by the value-added by inputs in each use. Since value-added from the supply of non-conventional output (environmental services) is accounted in the generalized system, either in the hypothetical scenario described above or in the actual situation in the Indian economy, the ENP estimated using the Generalized System of National Accounts will be higher than GNP estimated using CSNA. Also in a hypothetical economy in which there are investments in pollution abatement, ENP estimated by treating environmental resources as primary inputs and valuing them at their shadow prices will be higher than GNP estimated using CSNA.

## 8.5 CONCLUSION

Planning for sustainable development requires an estimate of environmentally-adjusted GNP (EGNP). The choice between preservation and developmental uses of environmental resources can be made so that EGNP is maximized. The conventional system of national accounts takes into account the supply of all goods and services for which markets are present, including some goods and services supplied by environmental resources. It is only goods and services for which markets are absent like amenity services, biodiversity, soil conservation, waste-disposal services, etc. that pose formidable problems in the estimation of EGNP.

EGNP can be estimated using a set of shadow prices for all the primary resources including the environmental resources. Alternatively, it can be estimated using the expenditure method described in Dasgupta and Maler (1994) and in Section 8.2. Also, the production method can be used for measuring ENP as described in Section 8.3. The production approach for measuring ENP outlined in this paper requires the definition of the control region for an environmental resource for internalizing the externalities and public goods type of services supplied through its preservation. However, it has to be recognized that while it is easy to define control regions for water pollution, it is difficult to do so for atmospheric or air pollution.

The United Nations' System of Integrated Environmental and Economic Accounting (SEEA) deals with both the concepts and measurement of sustainable income and the valuation of environmental resources. Using the maintenance valuation approach of SEEA, an attempt is made to estimate the cost of water pollution abatement for the Indian industry. Using data from a survey of 17 highly water-polluting industries and data from the Annual Survey of Industries, estimates of capital stock, primary input cost and total annual cost for water pollution abatement are obtained for a hypothetical situation in the indian economy in which all the factories in an industry invest in water pollution abatement. However, in reality, investments undertaken by industries for pollution abatement is very nominal as revealed by a FICCI survey, not even 25 per cent of factories are meeting pollution standards. ENP should take into account the value-added from the net supply of environmental resources. Since the generalized system of National Accounts follows this procedure, ENP calculated by this method will be higher than GNP estimated using CSNA.

# 9. Valuation of Benefits to Households from Industrial Water Pollution Abatement Practices at Nandesari[1]

SMITA MISRA

Estimation of benefits associated with costs of water pollution abatement is a difficult problem. The basic idea is to place values on environmental regulations involving water quality preservation. While the scientific rationale behind water quality preservation is well understood, its economic rationale for a developing economy has to be verified. This paper focuses on the feasibility of measuring benefits of water quality regulations. This depends on whether peoples' preferences are well-formed to place values on the effects of environmental regulations as in the case of water quality preservation. This paper attempts to tackle these issues with the objective of translating the benefits from water quality preservation into monetary values. A case study of the Nandesari Industrial Estate in the Gujarat state of India is taken up for this purpose.[2]

## 9.1 ECONOMIC VALUES OF ENVIRONMENTAL RESOURCES

The conventional theory of benefits measurement describes economic values of environmental resources based on the concepts of compensating and equivalent measures of welfare change. These measures are based on static utility maximization describing individual behaviour. Economic values are classified as user values and

[1] This chapter is part of my doctoral dissertation to be submitted to the University of Delhi.

[2] See Appendix A 9.1 for information about the Nandesari Industrial Estate.

non-user values. User values are defined as the benefits to direct users of the resource like recreation and other economic benefits. Krutilla (1967) brought into focus non-user values playing an important part in the valuation of environmental resources like 'water quality'. 'Existence value' recognizes that welfare of some individuals could increase simply with the knowledge that a resource exists. Existence values are defined in a framework that assumes certainty and do not require consumption of the resource. Either altruistic motives or vicarious consumption such as reading magazines, viewing films, listening to lectures about environmental resources can motivate the existence values. In addition, the consumer could have 'bequest value' for the resource, i.e. he attaches an economic value to the environmental good for the benefit of future generations.

A different perspective of welfare measurement introduces uncertainty into the description of consumer behaviour. When an individual is uncertain about his demand for an environmental good and is also faced with uncertainty about the availability of that good in the future, he would be willing to pay a sum over and above his expected consumer surplus to guarantee that the good is available for his future use. Weisbrod (1964) calls this the 'option value' of an environmental resource. The option price of a resource is defined as the option value plus the expected consumer surplus.

Thus, Total Economic Value (TEV) can be expressed as:

TEV = Use Value + Existence Values + Bequest Value + Option Value

This paper attempts an economic valuation of all these values for the benefits derived due to pollution abatement activities at Nandesari Industrial Estate.

Fig. 9.1 describes the conservation benefits for the villages surrounding Nandesari Industrial Estate and the urban area of Vadodara. It shows the user and non-user values derived from the preservation of water quality. While non-user benefits are the same for both the regions, user benefits differ. The rural consumers have more pronounced damages in terms of damages to health, productivity losses and time losses. The urban consumer has losses related to the source of drinking water. One third of Vadodara gets water from Mahisagar source, which is polluted. There could be losses to others if they eat toxic fruits and vegetables which are grown using polluted waters of Mini and Mahisagar rivers.

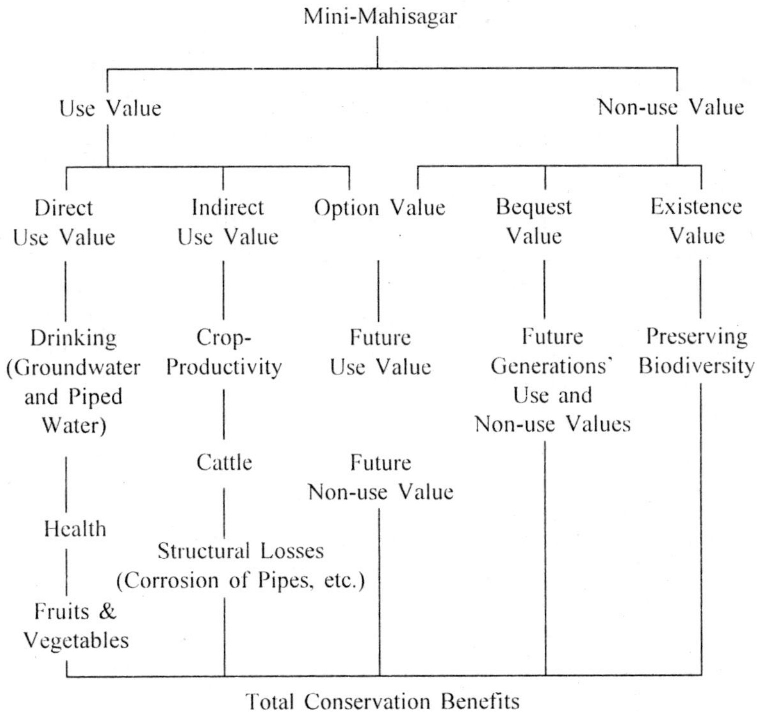


Fig. 9.1 Conservation Benefits

## 9.2 THE NEED FOR CONTINGENT VALUATION

The market fails to generate a true price of the environmental good 'water quality'. Hence, one cannot arrive at a cost of this amenity directly. Since water quality affects the utility of individuals, an economic value exists and the idea is to capture this 'economic value'. A method which could best capture the total economic value discussed in the last section for rural and urban areas surveyed should be used. The various approaches for the measurement of this economic value can be classified as:

(a) hedonic price approach—those based on direct and indirect market information, such as property values, wage rates, expenditure on related goods, etc.;

(b) physical linkage approach—those based on dose-response data linking environment changes to pollutants;

(c) market-based approach such as travel cost methods/contingent valuation—those based on stated preferences in the absence of well-defined markets, expressed through questionnaires, etc.

The hedonic price approach looks for a market in which goods are bought and sold and observes that environmental factors are attributes of those goods or factors. Different locations have varied environmental attributes which result in differences in property values. The property price effect due to differences in pollution levels can be obtained with the help of a multiple regression technique. This approach can be applied only when households are aware of the costs or benefits of the environmental attribute and are able to adjust their residential locations to choose whatever combinations of attributes they want. Hence, not only should the consumers be aware of the costs/benefits of the water quality, but they should be able to adjust their location to reveal their preferences. Markets should be functioning and the consumers should be well informed (Markandya, 1992; Freeman, 1994). This cannot be expected for the rural or the urban areas surveyed in this case study. Given the developing economy context, neither do the markets function well nor are the consumers well-informed. Further, this method does not consider the non-user values which may form a substantial portion of total economic value for the residents of rural and the urban areas.

People's preferences are not taken into account in estimating price for water quality using the physical linkage approach. A technical relationship is assumed between the consumer and water pollution. There are serious shortcomings of this approach. Some effects like 'productivity losses' are observable and market data would be available. However, there could be important effects which might not be observable in the market. Further, the physical linkage methods do not capture the non-user values which are very important in the case of Mini and Mahisagar rivers.

The market-based approaches use a direct approach. Travel cost models pay special attention to the value of time and the choice of site visits. They are used to value benefits of improvements in recreational facilities in parks, cultural sites, etc. The travel cost method does not apply to this case study as no recreational sites or parks are being valued. The contingent valuation method asks people what they are willing to pay for a benefit and/or what they are willing to receive by way of compensation to tolerate a cost if a market existed for the good in question. Given the limitations of physical linkage and revealed preference methods, this paper relies on the expressed preference method or the contingent valuation (CV) approach. Specific advantage of this method in this study would be the ability to measure existence, bequest and option values; providing a measure to value goods not currently available and to directly estimate the relevant ordinary and Hicksian demand curves.

In a recent debate on contingent valuation (Portney, 1994; Hanemann, 1994; and Diamond and Hausman, 1994), there are simplistic dichotomies at work. Surveys of attitudes could be fallible and much would depend on context and interpretation; stated preferences are undependable; outside the market, people are ignorant and confused; as consumers people can be taken seriously, as voters they cannot be taken seriously etc. Contingent valuation has been criticized for context effects, framing effects and embedding effects. The National Oceanic and Atmospheric Association Panel (NOAA, 1993) saw these as extreme arguments and rejected them. Contingent valuation provides a way of tracing the demand curve for a commodity that cannot be revealed through market data.

## 9.3 WELFARE MEASURES

### 9.3.1 Willingness to Pay Versus Willingness to Accept Formats

In the case of public goods like 'water quality', the individual is faced with a quantity rather than a price constraint, since the good is unpriced. Public goods like 'water quality' also have a much higher income elasticity than marketed goods. This is especially true in a developing economy context where 'water quality' is a luxury good which could be affordable only when adequate food, clothing and shelter have been acquired. Hence, the income effect due to a change in the 'water quality' provision may undermine the consumer surplus measure of welfare change. Hicksian compensating surplus (CS) and equivalent surplus (ES) measures could be used to provide welfare measures in the case of quantity constrained utility functions. Table 9.1 gives the kinds of welfare measures generally used.

For the urban survey, a willingness to pay (WTP) measure has been used for both the user and non-user values. This conforms to Type 1 of a proposed change in Table 9.1, measuring the compensating surplus for ensuring that the change occurs. For the rural survey, a willingness to accept (WTA) measure for welfare loss of Type 4 in Table 9.1 has been used to assess user values while a WTP measure for welfare gain of Type 1 in Table 9.1 has been used to assess non-user values.

Table 9.1
Welfare Measures

| Proposed Change | | Welfare Measure |
|---|---|---|
| 1. Welfare Gain | (CS) | Willingness to Pay to ensure that the change occurs. |
| 2. Welfare Gain | (ES) | Willingness to Accept if gain does not occur. |
| 3. Welfare Loss | (ES) | Willingness to Pay to avoid loss occurring. |
| 4. Welfare Loss | (CS) | Willingness to Accept if loss does occur. |

The choice of the elicitation method depends on the acceptance of WTP or WTA as the relevant mode of valuation in a specific case. Willig (1976) and Randall and Stoll (1980) show that WTP and WTA would be close in value, depending directly on the size of the income elasticity of demand for the commodity whose price changes. The literature maintains that it is implausible to have large discrepancies between WTP and WTA merely because of the implied property right in the case of WTA. Also loss aversion reflected in 'prospect theory' implies that value functions are concave for gains and convex for losses and steeper for losses (Kahneman and Tversky, 1979). Hanemann (1991) extended Willig's analysis to show that the disparity between WTP and WTA depends on the ratio of the income elasticity and an elasticity of substitution between the environmental good and all other goods. Larger ratios of WTP and WTA have arisen in cases of unique or irreplaceable environmental goods.

Given the large disparity between WTP and WTA in empirical studies, the 1993 NOAA panel recommends the use of WTP format. However, empirical analysis contradicts this view (Knetsch and Sinden, 1984; Kahneman, Knetsch and Thaler, 1990). Evidence indicates differences between valuations of gains and losses reflecting peoples' preferences. They value losses at a much higher level than gains of an equal amount. Hence in deriving welfare assessments, the indifference curves will not be independent of the initial entitlement and direction of exchange offers (Knetsch, 1989). There will be a kink in the indifference curve at the reference point

(Knetsch, 1994). Hence property rights linked to the choice of welfare is misleading. Actual changes in welfare would be linked to the initial point from where a gain or a loss is measured.

In the rural survey, WTA format is justified in view of the reasoning given above. Given irrigation, washing and drinking qualities of water, a kink in the indifference curve would occur when the quality of water deteriorates. Thus 'D' is the correct reference point in Fig. 9.2 since drinking water of very good quality was available in this area. When 'D' quality of water is withdrawn, the welfare loss would be measured by the sum necessary to leave the affected persons on the same indifference curve measured by the compensation demanded or WTA. Given a movement from 'D' to 'I', the WTA would be W1 W3 in Fig. 9.1. (For a movement from 'W' to 'I', the WTA would be W2 W3). The WTP would have been a correct measure if, 'I' or irrigation quality water was the reference point and a movement towards 'W' or 'D' would have been a gain. Hence welfare loss in terms of WTA when a loss has occurred would be the appropriate measure for the demand for environmental good 'water quality' which is a 'luxury' good for low income, uneducated respondants in a rural setting.

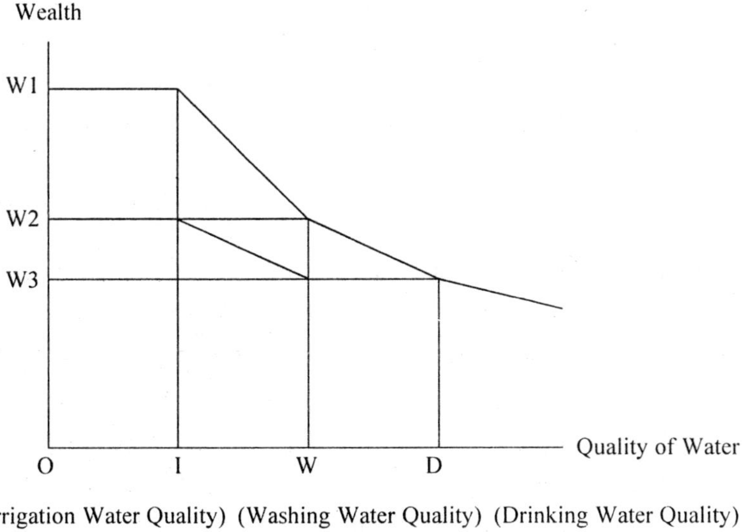


Fig. 9.2 Changing Preferences of Households with Water Quality Changes

NOAA (1993) supports WTP as the correct format, simply because it is the lower bound. Following this argument, if we use WTP for the rural survey, there would be an undervaluation of the 'damages'.

### 9.3.2 Welfare Measures and Demand for Water Quality

Let us consider the household utility function for the affected households in the Nandesari Industrial Area as

$$U^h (X^h, Q), \quad h = 1,2,...H \tag{9.1}$$

where $X^h$ is a $N x$ 1 vector of private goods consumed by the $h$th household and Q is the quality of water, the public good. Assuming that the quality of water is an unpriced pure public good, the budget constraint of the $h$th household is given as

$$P X^h = M \tag{9.2}$$

where M is money income. From (9.1) and (9.2) we have conditional demand functions (conditional upon the imposed public good Q) for private goods as

$$X_i = X_i (P, M, Q), \quad i = 1, 2, 3, \ldots, N \tag{9.3}$$

where P is a NX1 vector of prices of private goods. Inserting (9.3) in (9.1) gives the conditional indirect utility function

$$V = V (P, M, Q) \tag{9.4}$$

Inverting (9.4) for M gives a conditional expenditure function

$$e = e (P, Q, U) \tag{9.5}$$

The dual to the utility maximization problem can be stated as: Minimize

$$e (P, Q, U) \tag{9.6}$$

subject to $U \geq U^0$ The solution to (9.6) gives the restricted expenditure function

$$e (P, Q, U^0) \tag{9.7}$$

The marginal value of a small increase in the quality of water is given as

$$W_Q = \frac{\delta e (P, Q, U^0)}{\delta Q} \tag{9.8}$$

where W is interpreted as the reduction in income that is just sufficient to maintain utility at the level corresponding to no improvement in the quality of Q.

The compensating surplus (CS) and the equivalent surplus (ES) measures for measuring the welfare change due to a change in the quality of water can be defined as follows. CS is the solution to

$$V\,(P, M, Q^0) = V\,(P, M-CS,\, Q^1) \text{ where } Q^1 \geq Q^0 \qquad (9.9)$$

while ES is the solution to

$$V\,(P, M+ES\; Q^0) = V\,(P, M, Q^1) \text{ where } Q^1 \geq Q^0 \qquad (9.10)$$

Using expenditure functions, CS and ES are defined as follows:

$$CS = e\,(P, Q^1, U^1) - e\,(P, Q^0, U^1) = M - e\,(P, Q^0, U^1) \quad (9.11)$$

$$ES = e\,(P, Q^l, U^0) - e\,(P, Q^1, U^1) = e\,(P, Q^1, U^0) - M \quad (9.12)$$

In the contingent valuation methods used in this study, willingness to accept compensation by the rural households for the damages they suffer due to a fall in the water quality, or the willingness to pay by the urban households for the improved water quality can be measured by either CS or ES measures. For instance, willingness to accept of *h*th rural household can be measured as

$$WTA^h = e\,(P, Q^1, U^1, S^h) - e\,(P, Q^0, U^1, S^h) + \varepsilon^h{}_1 - \varepsilon^h{}_0 \quad (9.13)$$

where $S^h$ is a vector of socio-economic characteristics of the *h*th household; and

$\varepsilon_0^h$ and $\varepsilon_1^h$ are error terms pertaining to expenditure functions of the *h*th household.

Equation (9.13) can be alternatively written as

$$WTA^h = I^h - e\,(P, Q^0, U^1, S^h) + \varepsilon^h \qquad (9.14)$$

where $I^h$ is the income of the *h*th household with unpolluted water and

$$\varepsilon^h = \varepsilon_1^h - \varepsilon_0^h$$

$WTA^h$ is defined as the difference in the minimum expenditures required for the rural household to be as well off with the polluted water as it was with unpolluted water. The random component of $WTA^h$ is a result of incomplete information about the household expenditure function from the observed data. $WTA^h$ is specified as a random variable which is a function of variables like household income, age and sex composition, education and other socio-economic variables and dummy variables reflecting household's environmental awareness, nature of work etc. The $WTA^h$ function can be written as

$$WTA^h = X^h \beta + \varepsilon^h \text{ where } \varepsilon^h \sim N(0, \sigma^2) \tag{9.15}$$

and where $X^h$ and $\beta$ denote respectively the vector of explanatory variables and parameters.

## 9.4 DESIGN OF THE QUESTIONNAIRE FOR THE CONTINGENT VALUATION SURVEY

### 12.4.1 National Oceanic and Atmospheric Association Recommendations

Some of the NOAA Panel (1993) guidelines for a valid contingent valuation study are:

- personal interviews
- careful pre-testing
- WTP format for elicitation
- referendum format
- reminder of substitutes and income constraints
- accurate description of the provision of the good
- yes/no follow-ups to ascertain why respondents answered yes or no to a WTP question
- cross-tabulations to interpret WTP in terms of attitudes towards environment.
- cross-checks on understanding and acceptance
- response to WTP should not be a deflection of 'warm glow' serving the same function as charitable contributions but should reflect the economic value of the environmental good.

The questionnaire used for this paper while adopting most of the recommendations of the NOAA panel, has been suitably adapted to elicit values in a developing economy context.

### 9.4.2 Focus Groups and Pre-testing

The questionnaire was tested through a series of focus group interviews conducted in the field. Two separate questionnaires for urban and rural areas were prepared for user and non-user values. The findings at the pre-testing stage were quite interesting. For the urban survey, non-user values were very evident, while for the rural survey, the user values were more important. In fact at the pre-testing stage, non-user values for the rural survey were quite disappointing. The villagers did not want to elicit a sum for non-user values, as they did not care for the future and did not hold a philanthropic view about the quality of life. This was expected from rural areas with low incomes and lack of education. Consequently, the final survey questionnaire was well-equipped with special cards giving information about preservation of biodiversity, and so on. The questionnaire for the pre-testing of the rural survey included both WTP and WTA questions for the values. Due to large protest bids (90 per cent) of the sample surveyed, the WTP format for user values of rural survey was dropped (confirming Knetsch, 1994 study). Detailed information in the form of cards was incorporated to be given to the respondents in the final survey questionnaire relating to user and non-user values.

### 9.4.3 The Survey Instrument

The Survey consists of two separate areas, an urban sample of Vadodara city and a rural sample of six villages surrounding Nandesari. These areas have been identified as affected regions due to the pollution of rivers Mahi and Mini. The kind of damages suffered in the two areas are very different. In the urban sample, the damages are related to drinking water and are not severe. On the other hand, in the rural sample, the damages are related to the livelihoods of the respondents as well as health damages. Thus two separate questionnaires relating to the respective damages were designed.

#### *Urban Questionnaire*

The first section related to socio-economic characteristics. Detailed information was obtained regarding the age, education, nature of work, family size and income, etc. The second section related to user values, attitudes, preferences and value elicitation for user benefits. The information regarding the quality of water was in the form of short questions. Special cards were used by enumerators to define the various kinds of user values the respondents would enjoy from better 'quality of water'. The extent of losses suffered by the respondents were elicited in monetary terms just preceding the WTP question on user values. Hence a realistic, meaningful and plausible scenario was created for a hypothetical market. The elicitation question for user values was in terms of WTP contributions to a trustworthy charitable organization. The pre-test showed that the people did not have much trust in the local government and would prefer a trustworthy organization for the 'clean-up'.

The obvious way to reveal the economic value of water quality is to ask respondents directly for their WTP or WTA value and record the stated amounts. This is the direct open-ended question method. A clear advantage of this method is that no clues are given in the questionnaire that may imply the economic value of water quality. The iterative format helps the respondents to consider their preferences carefully as the respondents are asked to reply to increases in WTP till they answer 'no' and the final WTP is recorded. In this case, an open-ended WTP question was asked with an iterative format, until they answered 'no'. This is again a deviation from the routine closed-ended referendum types of questions asked in the literature. The user values would differ from individual to individual, depending on their actual losses. A monetary value of actual losses with respect to crop productivity and treatment costs due to polluted water preceded the elicitation question. Hence, the respondents could arrive at their WTP value more accurately. Finally, if the answer was 'do not know', the respondent was asked for an explanation.

The non-user part of the urban questionnaire, first provided information of non-user values that the respondent could derive from better quality water, substantiated by cards with the enumerator. Reminder about the income constraint and other uses for the income was provided. The WTP for non-user values was an open-ended iterative format question. If the answer was 'do not know', the respondent was asked to explain why. The last section of the questionnaire consisted of 'follow up' questions to verify the attitudes and preferences of the respondent with respect to environmental awareness, purchases of equipment to purify water and belief in bequest and existence values. Finally, the declaration of the interviewer about the length of the interview and the understanding of the respondent was taken.

*Rural Questionnaire:* The first part consisted of socio-economic characteristics with details about age, education, nature of work, family size, gross family earnings, family assets, and cropping pattern.

The second part consisted of user benefits in terms of actual damages to the respondent. Information regarding the Nandesari industries and the level of water pollution was provided to the respondent. A special card was developed providing information about the state of water quality in 2015 if the current trend of pollution continued and the state of water quality in 2015 if pollution abatement strategies were followed. Specific questions relating to the type of damages and actual monetary values for the damages were asked. Reminders about the benefits from Nandesari industries in terms of employment and income were given. The WTA question was elicited for the total damages they suffer in a year. This question used a closed ended referendum format. Three levels were given as a starting point, Rs 5,000 for farmers possessing no land, using 500 as iterative bids; Rs 10,000 for farmers possessing between 1 and 5 acres of land using Rs 1,000 as iterative bids; Rs 20,000 for farmers possessing 6 or more acres of land using Rs 2,000 as iterative bids. Since WTA is an upper bound of the economic value, the respondent was asked to reply to the respective levels of iterative bids. The iterative bids are respective increases/decreases in the WTA amount offered. If the respondent said 'yes' to a particular bid level, the bid was decreased till he said 'no', he is not willing to accept and the last bid to which he agreed was recorded. Similarly, if the respondent said 'no' to a particular bid, the bid was increased till he answered 'yes' and the final bid recorded. The starting points were obtained in the pre-test survey, after interviewing sarpanches of 6 villages and assessing the damages. An explanation was asked for if the answer was 'do not know'.

For the non-user section, detailed information about non-user values was provided with the help of cards. An open-ended WTP format with an iterative bid was used till they answered 'no'. Both the initial bid and the final bid were noted.

Follow-up questions were asked to confirm the preferences and attitudes of the villagers. Questions relating to environmental awareness were asked. Finally, a declaration by the interviewer was taken.

The urban and the rural surveys were conducted by twenty-two college students. These boys (all residents of Vadodara) were trained for the contingent valuation survey for three days, before they undertook the final survey.

### 9.4.4 Sample Information

*Urban Survey:* The city of Vadodara was divided into two broad zones according to (1). Mahi river as a source and (2). Sayaji Sarovar (Ajwa) as a source of drinking water. The former is a polluted source while the latter is an unpolluted source. Random sampling was undertaken to cover high income, middle income and low income areas in each of these areas. The following areas were covered:

Mahi river as source of water:

1. High Income : Alkapuri, Akota, Producivity Road, Tandleja, Race Course, Saydvasna Road, Wadiwadi Area.
2. Middle Income : Gotri, Subhanpura, Fatehganj, Chhani
3. Low Income : Gonwa, Akota village, Gotri village, Tandleja village, Attadara village, Saydvasna village, Wadiwadi village.

Sayaji Sarovar (Ajwa) as source of water:

1. High Income : Ajwa Road, Waghodiya Road
2. Middle Income : Marni, Karlibag, All city area, Wadi, Baroda Dairy, ONGC
3. Low Income : Marni village, Kishanwadi, Tarsali village, Warashiya

Mahi river (the polluted source) supplies about 119 million litres per day (mld) of water to Vadodara city, compared to about 45 mld of water from Sayaji Sarovar (unpolluted source).

*Rural Survey*: Random sampling procedures were adopted to cover six villages—Angadh, Dhanora, Koyali, Sindhrot, Sherkhi and Nandesari. The choice of the villages was according to the proximity of these villages to Nandesari Industrial Estate and the extent of damage due to polluted waters of Mini and Mahisagar rivers. The sarpanch of each village was very helpful in identifying the extent of damages in the village.

## 9.5 ESTIMATES OF WTP FOR URBAN USER AND NON-USER BENEFITS

The variables used for regression are:

UUBID = Urban user bid values
UNBID = Urban non-user bid value

| | | |
|---|---|---|
| RUBID | = | Rural user bid value |
| RNUBID | = | Rural non-user bid value |
| PCE | = | Per capita earnings of the household |
| AGE | = | Age of the respondent |
| ED12 | = | Education greater than class 12 dummy |
| FNO | = | Number of family members |
| AWR | = | Awareness dummy |
| IS | = | Ideal solution dummy |
| EQUIP | = | Equipment costs |
| MCON | = | Membership of conservation groups dummy |
| QWLD | = | Quality of drinking water dummy |
| SSP | = | Structural problems dummy |
| DMH | = | Damages to health costs |
| IWK | = | Employment in industry dummy |
| RSSP | = | Interaction of rich dummy with SSP dummy |
| RDMH | = | Interaction of rich dummy with DMH dummy |
| ENIMP | = | Environment importance dummy |
| NAR | = | Nandesari responsibility dummy |
| ED9 | = | Education greater than class 9 dummy |
| NWK | = | Employment in industry dummy |
| DT | = | Damages due to time losses dummy |
| DH | = | Damages due to health losses dummy |
| DC | = | Damages due to crop losses dummy |
| DK | = | Damages due to cattle losses dummy |
| NEAR | = | Earnings from Nandesari factories |
| EN | = | Surrounding environment damages dummy |
| DMH | = | Damages to health costs |
| DMC | = | Damages to crop costs |
| ED6 | = | Education greater than class 6 dummy |

Specific questions relating to environmental awareness were asked. If the respondent is environmentally aware, AWR dummy is one, otherwise zero. If the respondent would like polluting factories to install pollution abatement plants and control pollution IS dummy is one and if he prefers the polluting factories to shut down altogether IS is zero. EQUIP is the amount of money (Rs in thousands) the respondent has spent on equipments like filters to purify water. If the respondent is a member of some conservation group MCON dummy is one, otherwise zero. If the respondent is satisfied with the quality of drinking water QWLD dummy is one, otherwise zero. If the respondent is facing structural problems like corrosion of pipes and loss of building materials due to polluted water, SSP dummy is one, otherwise zero. RSSP dummy is RICH* SSP where RICH dummy is defined for households total earnings greater than Rs 200,000 per annum. If the respondent/household is suffering from water quality related diseases, then DMH is the amount of money spent per annum for treatment. RDMH is RICH* DMH where RICH dummy is defined for households' total earnings greater than Rs 200,000 per annum. If the respondent is working in industries, then IWK dummy is equal to one, otherwise zero.

The results in Table 9.2 show that UUBID for urban households significantly depends on per capita earnings, age, family size, awareness, ideal solution, equipment, membership of conservation groups, damages to health, and damages to health of the rich people. Splitting the sample of households according to a RICH dummy was done to allow for the possibility that structural problem and health damages may have differential effect on WTP of rich and poor households. The use of this dummy gives interesting results. From the results, it may be inferred that the rich households can correctly relate the treatment costs to WTP for user values. An important explanatory variable is the defensive expenditures on equipment for cleaning drinking water.

### 9.5.1 Non-user Benefits of Urban Population

The results of some of the estimations for WTP of non-user values based on OLS method are presented in Table 9.3 below. The coefficients of R4 (regression) are used for estimating the WTP for non-user benefits.

ENIMP is a dummy for the importance the respondent attaches to the environment. If the respondent feels that the Nandesari factories should undertake measures to clean up the rivers, NAR dummy is one, otherwise zero. The other variables are as explained for Table 9.2 above.

The results show that the UNBID for urban households significantly depends on per capita incomes, age, higher education (above school level), size of the family, the responsibility they attach to Nandesari factories for polluting the environment, ideal solution and the quality of drinking water. Given the concept of non-user values in terms of existence and bequest benefits that the respondent would

Table 9.2

Results of WTP (User Values) Regressions for the Urban Area

Dependent variable log UUBID — No. of observations = 386

| Explanatory Variables | R1 | R2 | R3 | R4 |
|---|---|---|---|---|
| log PCE | 0.2085 (4.257)*** | 0.2083 (4.268)*** | 0.1828 (3.784)*** | 0.1548 (3.198)*** |
| log AGE | −0.2737 (−2.458)*** | −0.2730 (−2.464)*** | −0.2681 (−2.447)*** | −0.3070 (−2.611)*** |
| ED12 | 0.07492 (1.214) | 0.07526 (1.215) | 0.08453 (1.366) | 0.07199 (1.182) |
| Log FNO | 0.4939 (5.594)*** | 0.4937 (5.623)*** | 0.4859 (5.551)*** | 0.5197 (6.231)*** |
| AWR | 0.2161 (3.588)*** | 0.2169 (3.655)*** | 0.2320 (3.986)*** | 0.2276 (3.945)*** |
| IS | 0.2347 (3.786)*** | 0.2342 (3.805)*** | 0.2527 (4.115)*** | 0.2438 (4.109)*** |
| EQUIP | 0.0819 (4.110)*** | 0.08203 (4.122)*** | 0.07965 (4.100)*** | 0.08232 (4.395)*** |
| MCON | 0.1790 (1.399) | 0.1773 (1.404) | 0.2109 (1.639)* | 0.2850 (2.090)** |
| QWLD | −0.0395 (−0.468) | −0.0399 (−0.470) | −0.0803 (−0.939) | −0.0900 (−1.223) |
| SSP | 0.0074 (0.036) | 0.0073 (0.036) | −0.0670 (−0.325) | −0.0163 (−0.095) |
| DMH | −0.1540 (−1.071) | −0.1551 (−1.086) | −0.2171 (−1.664)* | −0.2437 (−2.229)** |
| IWK | −0.0065 (−0.105) | | | |
| RSSP | | | 0.5566 (1.135) | 0.5497 (1.114) |
| RDMH | | | 0.8639 (1.842)* | 0.8928 (1.862)* |
| CONSTANT | 3.3572 (6.141)*** | 3.3552 (6.133)*** | 3.6149 (6.795)*** | 4.0294 (6.858)*** |
| $R^2$ | 0.23 | 0.23 | 0.24 | |
| Adjusted $R^2$ | 0.20 | 0.20 | 0.22 | |
| F | 9.385 | 10.264 | 9.525 | |
| Log Likelihood | | | | −297.18 |
| Restricted Log L | | | | −306.83 |
| Chi-squared (13) | | | | 19.306 |
| Significance Level | | | | 0.11392 |

*Notes*: (t-values in parentheses); *** = significant at 1 per cent level; ** = significant at 5 per cent level; * = significant at 10 per cent level; Regression R4 is R3 corrected for heteroscedasticity using the Multiplicative Heteroscedastic Regression Model.

Table 9.3

Results of WTP (Non-user values) Regressions for the Urban Area

Dependent variable log UNBID — No. of observations = 366

| Explanatory Variables | R1 | R2 | R3 | R4 |
|---|---|---|---|---|
| log PCE | 0.15674 (2.417)*** | 0.15648 (2.403)*** | 0.15599 (2.467)*** | 0.10047 (1,847)* |
| log AGE | –0.54731 (–3.677)*** | –0.54268 (–3.668)*** | –0.53800 (–3.629)*** | –0.45826 (–3.366)*** |
| ED12 | 0.29831 (4.161)*** | 0.30120 (4.201)*** | 0.29177 (4.018)*** | 0.22473 (3.302)*** |
| log FNO | 0.53689 (4.663)*** | 0.53641 (4.673)*** | 0.53783 (4.689)*** | 0.43867 (4.495)*** |
| IWK | –0.03939 (–0.538) | | | |
| ENIMP | –0.09185 (–0.979) | –0.09506 (–1.018) | | |
| NAR | –0.49438 (–6.107)*** | –0.49394 (–6.098)*** | –0.49741 (–6.116)*** | –0.46036 (–6.748)*** |
| AWR | 0.18867 (2.442)*** | 0.19214 (2.528)*** | 0.19386 (2.536)*** | 0.09992 (1.472) |
| IS | 0.26527 (3.286)*** | 0.26255 (3.284)*** | 0.26603 (3.329)*** | 0.30462 (4.117)*** |
| MCON | 0.19137 (0.922) | 0.18154 (0.875) | | |
| QWLD | –0.16784 (–1.833)* | –0.16856 (–1.845)* | –0.15800 (–1.749)* | –0.21241 (–2.295)** |
| CONSTANT | 4.8268 (6.557)*** | 4.7988 (6.526)*** | 4.7696 (6.556)*** | 5.2455 (7.622)*** |
| $R^2$ | 0.29 | 0.28 | 0.28 | |
| Adjusted $R^2$ | 0.26 | 0.26 | 0.27 | |
| F | 12.831 | 14.112 | 17.474 | |
| Log likelihood | | | | –336.16 |
| Restricted Log-L | | | | –350.63 |
| Chi squared (B) | | | | 28.943 |
| Significance Level | | | | 0.000324 |

*Notes*: (t-values in parentheses); *** = significant at 1 per cent level; ** = significant at 5 per cent level; * = significant at 10 per cent level; Regression R4 is R3 corrected for heteroscedasticity using the Multiplicative Heteroscedastic Regression Model.

enjoy, income and education level are very important and the results support this understanding.

### 9.5.2 Estimates of Benefits to Urban Population

The per capita WTP (per annum) for user and non-user values has been estimated for different income classes of the urban sample in Table 9.4.

Table 9.4
Estimates of Per Capita WTP for User and Non-user Benefits for Urban Sample by Income Classes

| Income class (Rs) | WTP (User) | WTP (Non-user) |
|---|---|---|
| 0–30,000 | 64 | 52 |
| 30,000–60,000 | 76 | 58 |
| 60,000–90,000 | 82 | 61 |
| 90,000–120,000 | 87 | 63 |
| 120,000–150,000 | 90 | 65 |
| 150,000–180,000 | 93 | 66 |
| Mean income (38,586) | 74 | 57 |

*Source:* Estimated as explained in the text

## 9.6 ESTIMATES OF WTA FOR USER DAMAGES AND WTP FOR NON-USER BENEFITS OF RURAL POPULATION

### 9.6.1 User Benefits of Rural Population

The results of some estimations for WTP of user values using OLS are presented in Table 9.5. The coefficients of R4 are used for estimating the WTA for user benefits foregone.

If the respondent has acquired higher than class nine education, then ED9 dummy is one, otherwise zero. If the respondent is working in industries, then NWK dummy is one, otherwise zero. If the respondent has time losses in collecting water, DT dummy is one, otherwise zero. If the respondent has health losses, DH dummy is one, otherwise zero. If the respondent has crop losses, DC dummy is one, otherwise zero. If the respondent has cattle losses, DK dummy is one, otherwise zero. NEAR are the earnings (Rs in thousands) of the respondent from Nandesari factories. If the respondent has a

Table 9.5

Results of WTP (User Values) Regressions for the Rural Area

Dependent variable log RUBID — No. of observations = 405

| Explanatory Variables | R1 | R2 | R3 | R4 |
|---|---|---|---|---|
| log PCE | 0.17491 (3.232)*** | 0.065427 (1.198) | 0.061943 (1.149) | 0.095739 (2.235)** |
| log AGE | 0.071013 (0.740) | 0.023319 (0.244) | 0.029889 (0.311) | 0.11657 (1.166) |
| log FNO | 0.35219 (4.567)*** | 0.17628 (2.194)** | 0.17346 (2.173)** | 0.17928 (2.345)*** |
| ED9 | 0.23843 (2.443)*** | 0.17586 (1.872)* | 0.17301 (1.834)* | 0.20781 (2.458)*** |
| NWK | –0.36780 (–0.980) | –0.38567 (–1.219) | –0.40576 (–1.284) | –1.2244 (–6.210)*** |
| DT | 0.56824 (1.541) | 0.60245 (1.991)** | 0.66245 (2.069)** | 0.48185 (2.382)*** |
| DH | 0.61535 (1.683)* | | | |
| DC | 0.23958 (0.705) | | | |
| DK | –0.18714 (–0.922) | | –0.10884 (–0.597) | 0.014637 (0.103) |
| NEAR | –0.01571 (–2.414)*** | –0.0107 (–1.948)** | –0.0102 (–1.890)** | –0.00183 (–1.971)** |
| EN | –0.15103 (–0.983) | –0.13180 (0.904) | –0.13307 (0.928) | –0.26919 (–2.289)** |
| DMH | | 0.03122 (4.934)*** | 0.0315 (4.987)*** | 0.028140 (4.803)*** |
| DMC | | 0.01260 (5.252)*** | 0.0128 (5.3300)*** | 0.0105 (5.700)*** |
| CONSTANT | 6.0172 (7.794)*** | 7.6599 (10.766)*** | 7.7071 (10.935)*** | 7.2068 (12.187)*** |
| $R^2$ | 0.16 | 0.25 | 0.25 | |
| Adjusted $R^2$ | 0.14 | 0.23 | 0.23 | |
| F | 6.977 | 12.822 | 11.691 | |
| Log likelihood | | | | –417.18 |
| Restricted Log-L | | | | –430.85 |
| Chi squared (11) | | | | 27.344 |
| Significance Level | | | | 0.0040764 |

*Notes*: (t-values in parentheses); *** = significant at 1 per cent level; ** = significant at 5 per cent level; * = significant at 10 per cent level; Regression R4 is R3 corrected for heteroscedasticity using the Multiplicative Heteroscedastic Regression Model.

clean environment around his house, EN dummy is one, otherwise zero. DMH are damages to health in terms of money value of treatment costs (Rs in thousands) per annum. DMC are damages due to crop losses (Rs in thousands) per annum.

The results show that the RUBID for rural households significantly depends on per capita earnings, number of family members, education (above class nine), whether working in industries or not, damages due to time losses in collecting water, earnings from Nandesari factories, whether respondent has a clean environment around his house, treatment costs due to health damages, and losses due to crop damages. The employment in industry is negatively related to RUBID. This explains the attitude of the villagers. If industry is their source of livelihood, they will not attribute losses due to industrial effluent and will have a low WTA. Similarly, the greater the Nandesari earnings, the less are they willing to accept. Health damage costs and crop productivity losses are two most important factors determining the WTA value. The time lost in collecting water for domestic purposes as well as for irrigation purposes also turns out to be a significant factor in explaining WTA values given by the rural respondents. Also, the worse the state of environment surrounding the respondent's house, the more is his willingness to accept.

### 9.6.2 Non-user Benefits to Rural Population

The results of some estimations for WTP of non-user values, based on OLS method are presented in Table 9.6. The coefficients of R3 are used for estimating the WTP for non-user benefits by the rural households of Vadodara.

If the respondent has acquired higher than class six education, then ED6 dummy is one, otherwise zero. The other variables are as before.

The results show that the RNUBID for rural households depends on per capita earnings and education. The results support the basic understanding that incomes and education play an important role in forming non-user values. An important point here is that for rural areas, 'water quality' is a luxury good for respondents with low incomes and lack of education. The regression results give a low explanatory power of independent variables as indicated by a low $R^2$ and hence their lower significance. Thus the non-user values are not considered for assessing the welfare losses to the rural area for this study.

Table 9.6

Results of WTP (Non-user Values) Regressions for the Rural Area

Dependent variable log RNUBID — No. of observations 191

| Explanatory Variables | R1 | R2 | R3 |
|---|---|---|---|
| log PCE | 0.10355 (2.256)** | 0.098705 (2.240)** | 0.080417 (1.751)* |
| log AGE | 0.16516 (1.664)* | 0.16422 (1.665)* | 0.19022 (1.502) |
| log FNO | 0.051197 (0.689) | 0.045747 (0.626) | 0.0057215 (0.070) |
| ED6 | 0.14235 (1.854)* | 0.14177 (1.844)* | 0.13045 (1.706)* |
| EN | 0.013044 (0.131) | –0.011511 (–0.132) | –0.025383 (–0.308) |
| NEAR | –0.0019187 (–0.514) | | |
| CONSTANT | 3.2775 (6.109)*** | 3.3262 (6.444)*** | 3.4461 (5.992)*** |
| $R^2$ | 0.07 | 0.06 | |
| Adjusted $R^2$ | 0.03 | 0.04 | |
| F | 2.141344 | 2.552961 | |
| Log Likelihood | | | –138.51 |
| Restricted Log-L | | | –144.72 |
| Chi Squared (5) | | | 12.422 |
| Significance Level | | | 0.029436 |

*Notes*: (t-values in parentheses); *** = significant at 1 per cent level; ** = significant at 5 per cent level; * = significant at 10 per cent level; Regression R3 is R2 corrected for heteroscedasticity using the Multiplicative Heteroscedastic Regression Model.

### 9.6.3 Benefits to Rural Population

The per capita WTA (per annum) for user values has been estimated for different income classes of the rural sample in Table 9.7.

Table 9.7
Estimates of Per Capita WTA User Values of Rural Sample by Income Classes

| Income class (Rs) | WTA User |
|---|---|
| 0–3,000 | 2,453 |
| 3,000–6,000 | 2,725 |
| 6,000–9,000 | 2,861 |
| 9,000–12,000 | 2,955 |
| 12,000–15,000 | 3,027 |
| 15,000–18,000 | 3,086 |
| Mean income (4,235) | 2,709 |

*Source:* Estimated as explained in the text

## 9.7 ESTIMATION OF BENEFITS OF WATER POLLUTION ABATEMENT AT NANDESARI

### 9.7.1 Urban Population

The urban survey covered 386 households for the user values and 366 households for the non-user values (after rejecting outlier and protest bids), giving a WTP (user) estimate of Rs 74 per capita per annum and WTP (non-user) estimate of Rs 57 per capita per annum. The water pollution abatement benefits for Vadodara city is given in Table 9.8 using sample data.

Table 9.8
Estimates of Household (per annum) WTP for User and Non-user Benefits of Urban Sample and Population (Rupees Million)

| Per Annum | Number of Households | WTP (user) | Number of Households | WTP (non-user) |
|---|---|---|---|---|
| Sample HH | 386 | 0.086 | 366 | 0.063 |
| Population HH | 567,370 | 126.31 | 567,370 | 97.18 |

*Note:* Estimated as explained in the text; Data source for population figures is CMIE, Profiles of Districts Nov. 1993.

### 9.7.2 Rural Population

The rural survey covered six villages having a population of 7,890 households. The sample covered 405 households for the user values; giving an estimate of per annum per capita WTA equivalent to Rs 2,709 for user values. The water pollution abatement benefits foregone for the rural population is given in Table 9.9.

Table 9.9
Estimates of Household (Per Annum) WTP for User Values of Rural Sample and Population (Rs Million)

| Per Annum | Number of Households | WTA (user) |
|---|---|---|
| Sample HH | 405 | 5.49 |
| Population | 7,890 | 106.87 |

*Note:* Estimated as explained in the text; the data source for the population figures of the six villages is village/town Primary Census, Abstract, National Informatics Centre, Office of the District Collector, Ahmedabad.

## 9.8 RELIABILITY, VALIDITY AND BIASES

The reliability, validity and biases in contingent valuation method depends on the 'Theoretical Construct'. The 'Theoretical Construct' is the maximum amount of money the respondent would pay for the public good like 'water quality' or the minimum amount they would accept for the absence of desired 'water quality', if the appropriate market for that public good existed (Mitchell and Carson, 1989). The theoretical construct depends on the following scenario—the good, the method, the level of provision, the elicitation framework, and the payment vehicle.

The prime concern in this paper is to create a realistic, meaningful and plausible scenario, bringing the elicited WTP/WTA values as close to the real values as possible. While the urban survey uses WTP as the method for user and non-user values, the rural survey uses WTA for damages and WTP for non-user benefits. In the urban areas the pollution abatement programmes could actually bring about increases in welfare. In the rural areas, given the gravity of the situation and widespread damages to soil productivity, aquifers, health, and flora and fauna of the region, WTA could capture the real losses. In a developing economy context, the people might not have sufficient faith in the government machinery to implement the desired pollution abatement measures. Hence an alternative arrangement in terms of a charitable organization capable of implementing pollution abatement measures is supposed to provide the 'good'. A contribution to this organization is requested for bringing about water quality improvements.

The conditions regarding familiarity and plausibility of the scenario are met for both the urban and the rural surveys. Urban Vadodara dwellers are very much aware of the damages the industrial effluent is causing. There are various NGOs like the SO CLEAN (Society for Clean Environment) which are actively involved in filing petitions against the Nandesari industries, monitoring pollution abatement measures at Nandesari, etc. The NGOs have been successful in closing 26 of these factories which were not able to treat their effluents. For the rural areas the respondents have Mahisagar as the source of water supply and Mini as its tributary. There are widespread damages due to fertility losses of the soil, cropping pattern changes, health losses, etc. These are widespread losses and affect the livelihood of the residents. All this ensures that the scenario is meaningful, realistic and plausible.

The researcher could expect a large variance in WTP/WTA elicitation due to large 'true' variance in heterogenous population in contrast to homogeneous groups (Mitchell and Carson, 1989). This is typically true for a population of a developing country like India. In this survey, stratification has increased the efficiency of sample statistics. In the urban survey, care was taken to include high income, middle income and low income earners. In the rural survey, stratified sampling is on the basis of big, medium and small landowners. The initial bid for WTA elicitation is on the basis of landholdings with Rs 5,000 for farmers possessing no land, using 500 as iterative bids; Rs 10,000 for farmers possessing between 1 and 5 acres of land using 1000 as iterative bids; Rs 20,000 for farmers possessing 6 or more acres of land using Rs 2000 as iterative bids. The starting points were obtained in the pre-test survey, after interviewing sarpanches of 6 villages and assessing the damages.

There are various 'biases' which have to be minimized from our understanding of the respondents' behaviour, to bring the observed value closer to the real value.

### Strategic Bias

These arise when respondents like to influence the outcome to serve their personal interests. This has been taken care of both in the urban and the rural surveys. In the urban survey, a charitable organization is supposed to bring about water pollution abatement. A caveat is provided to the respondents that if WTP is not sufficient, the organization might not be able to preserve the water quality to the desired standards and the purpose of such a programme will be defeated. Similarly in the rural survey, strategic bias is kept to the minimum by reminding the respondents that WTA should be minimum, and close to the actual losses suffered, otherwise it would be difficult to undertake any programme relating to reduction of these losses.

### Hypothetical Bias

This is taken care of by providing a meaningful, realistic and plausible scenario to the respondent. Given the gravity of the situation due to the pollution loads of Mini and Mahisagar rivers, the hypothetical construct truly depicts the picture.

### Starting Point Bias

This is taken care of by eliciting open ended WTP questions in both WTP (user and non-user, urban) and WTP (non-user, rural). Initial bids are provided in WTA (rural) elicitation according to their landholding. (The pre-testing showed losses related to the size of landholdings in the rural areas). An iterative procedure is then used to get the maximum WTP and minimum WTA values.

### Scenario Mis-specification Bias

There could be a 'scenario mis-specification bias', relating to the good, type of market, payment vehicle, elicitation question or embedding effect. These have been taken care of in the design of the questionnaire. The good, that is 'water quality' is the main problem in this area. Hence the respondent is already aware of 'the good' and its associated effects. For the urban survey a WTP for both user and non-user benefits has been used. The respondent can link the provision of the public good 'water quality' with its associated damages. In this sense, the respondent will have a value attached to the 'water quality' and a private good market like a charitable organization providing the 'good' will work. In the developing economy context, with an over-taxed consumer, a tax for the improvement of 'water quality' will not work. Hence charitable contributions are asked, keeping in mind that water quality will improve not just for the respondent, but for fellow citizens too. Reminders are given to the respondent, that since it would be a huge task to preserve 'water quality' to standards prescribed by the Pollution Control Board, any free riding will not help. Given the immense problem, with respect to 'water quality' in this region the private market and the contribution to a trustworthy charitable organization as a payment vehicle worked.

In the context of the rural survey a WTA format is used for user value damages and a WTP format is used for non-user values. A referendum type market is used for the WTA format. It provides initial starting points for the damages to the respondents. The starting values are deliberately kept low (according to data from the pre-test survey). An iterative procedure is then used to get the final values. In this way the respondent's tendency to overstate WTA is taken care of. Since the respondents do not trust the government (it has been unable to clean the rivers in the last 25 years), contributions to a trustworthy charitable organization are desired. A WTP format is used for the non-user values. Since positive non-user benefits would be received by the respondent if the water quality is preserved, the private market context with contributions to a charitable organization would work. Since values (user and non-user) are elicited for rivers Mini and Mahisagar, the benefits from preservation would be related to these resources only. Hence the question of an embedding effect will not arise.

## 9.9 CONCLUSION

The results of this study show that it is possible to evaluate meaningfully the benefits of regulatory practices related to water pollution abatement at Nandesari Industrial Estate. Well-designed and carefully administered Contingent Valuation Surveys adapted to the requirement of the actual situation in the developing countries is necessary. It might however require some deviation from the standard practice followed in the developed countries.

The empirical results show that predominant user and non-user values do exist even among the relatively uneducated and low income groups in India. The important question is how to estimate these latent values. The urban survey conducted for Vadodara city and the rural survey conducted for six villages surrounding Nandesari Industrial Estate are attempts to estimate these values.

The urban WTP user survey covered 386 households giving an estimate of Rs 74 per capita (per annum) and urban WTP non-user survey covered 366 households giving an estimate of Rs 57 per capita (per annum). The total WTP for user values and non-user values estimated for urban Vadodara are Rs 126.31 million and Rs 97.18 million per annum respectively. The results show that WTP for user values by urban households significantly depends on per capita earnings, age, size of family, awareness, ideal solution of polluters treating the effluents, expenditure on equipment to clean water like filters etc., membership of conservation groups and damages to health of the richer section of the society. The WTP for non-user values by urban households significantly depends on per capita earnings, age, whether or not they have higher education (above school level), the size of the family, the responsibility they attach to Nandesari for pollution, level of awareness, and the ideal solution of Nandesari adopting pollution abatement strategies rather than shutting down the factories.

The user value of the urban sample is very different from the user value of the rural sample.[3] This is because the damages in the urban area are restricted to consumption of drinking water whereas the damages in the rural sample are related to their livelihood as well as health. The rural survey covered six villages having a population of 7,890 households. The sample covered 405 households giving a per capita per annum WTA (user) estimate of Rs 2,709. The total WTA for user values in rural area surveyed is Rs 106.87 million per annum. The results show that the WTA damages by the rural population significantly depend upon the per capita earnings, size of the family, education level greater than class nine, employment in industries, time losses in collecting water, earnings from Nandesari, state of the surrounding environment, costs of health and crop damage.

[3] An estimate for actual losses of the urban area and rural area has been made using the sample information. For the urban area, Rs 220 per capita (per annum) are the defensive expenditures on equipment like filters and aquaguards. This gives a loss of Rs 375 million for the urban area in this study. For the rural area, Rs 3,040 per capita (per annum) is the reported cost due to crop productivity losses and treatment costs for health damages. This gives a loss of Rs 120 million for the rural area in this study.

An interesting result that emerges is that non-user values would form an important part of benefits for the educated, high income groups who are environmentally conscious. Hence there are significant factors explaining the non-user values in the urban sample. For the rural sample, the environmental good, 'water quality', is a luxury good and the non-user benefits they derive from it are insignificant and are not considered for assessing the welfare losses to the rural area in this study.

There are clear economic linkages between conservation efforts and the benefits derived from these efforts. The effluent control cost to Nandesari Industrial Estate is Rs 129 million per annum (Misra, 1996). The findings indicate that large potential welfare gains could be generated by water pollution abatement practices at Nandesari and underscore the need to undertake the necessary abatement measures.

# APPENDIX 9

## Background History of Nandesari Industrial Estate

The first major chemical factory to be set up in Vadodara was the Universal Dyestuff Industry at Sankarda in 1960, about 10 km north of Vadodara on the banks of the river Mini, a tributary of river Mahi. The easy accessibility of these rivers served as a disposal bin and attracted other industries to this area.

The Mahi river originates in the Vindhya hills in Dhar district of Madhya Pradesh and flows through the state of Madhya Pradesh and Gujarat over a distance of 600 km (360 miles) before it joins the Arabian Sea at the Gulf of Cambay. The Mini river, a tributary of Mahi originates at a point about 30 km north-east of Vadodara. It courses parallel to Mahi and flows towards the west before rejoining Mahi between the villages of Jaspur and Sindhrot about 15 km downstream of Vasad. The first reports on pollution of the Mahi river and consequent damage to aquatic life came in October 1968. The inhabitants of Bajwa and Undera villages reported to the government about contamination of water in their village tanks and wells causing the death of some animals. These kind of complaints became a regular feature from this area. Vast areas of fertile land have been turned into totally barren wasteland. Effluents being dumped into Mahi and Mini rivers rendered them unsuitable for aquatic life and human consumption or use. Nearby areas of Nandesari, like Angadh, Sindhrot, Dhanora, Koyali and Sherkhi villages are faced with the problem of groundwater contamination as well. Cattle and fish death after drinking the water has occurred at the confluence point of Mini and Mahi. As a consequence of these reports regarding the increased pollution load of the Mahi and Mini rivers, the Government of Gujarat appointed a technical committee to look into the matter. The committee recommended an effluent channel to convey the industrial waste waters from Vadodara to Jambusar for discharge into the estuary at the Gulf of Cambay. A 56 km long effluent channel was built to carry the effluents to the Gulf of Cambay. The factories were required to treat the effluents in their own treatment plant before disposing them into the collection wells at Dhanora from which they are conveyed into the channel. This channel over the years has served as an excellent source of water for the farmers.

A recent study (Sharma, 1995), analyzes water quality with respect to both physical and chemical parameters laying emphasis on metals in estuary sediments and soil of adjoining areas. Metal content in vegetables, grains etc. and histopathological alterations in the tissues of a common fish were examined. This study reached alarming conclusions. The analysis of metal content in the estuary sediment shows very high values, which are much higher than the reported values of sea waters of India. The study shows relatively higher levels of all metals analyzed in the upstream area as compared to the downstream area. The reason ascribed to this are, the construction of a dam upstream making the tidal ingress carry back the metallic contaminants to the upstream area, and unchecked dumping of untreated effluents into the river Mini. The high metal concentration for nickel, lead and zinc at upstream points suggests discharge of untreated industrial effluents into the Mini. The Mahi estuary at the Gulf of Cambay reveals rapid erosion in the quality of estuarine flora and fauna. The study reveals alarmingly high levels of total solids, TDS, COD, chloride sulphate, nitrate and metals in the water of wells situated at about 50–200 m from the effluent channel.

The content of metals is much higher in fruits, vegetables, cereals and grains from the channel areas as compared to non-channel areas. This is not only toxic and hazardous for health, but implies a lower nutritive value as well. The population of Baroda who consume these vegetables and grains are continuously bio-accumulating these metals in their bodies which may cause various ailments. The increasing loads of pollutants in the effluent channel, uncontrolled pilferage of channel water for irrigation, together with continuous disposal of untreated effluents in Mini and Mahisagar will soon render the underground water unfit for human consumption and soil unfit for human subsistence.

Thus the water pollution of Mini and Mahisagar rivers and underground water has led to serious water and soil degradation around the effluent channel and human health hazards due to consumption of fruits/vegetables grown in this region. These serious impacts of effluents from Nandesari factories motivated this paper. The contingent valuation technique has been used to assess the user and non-user values for 'water quality' preservation from urban Vadodara and rural areas, comprising six villages around river Mini.

# 10. Accounting for Costs of Water Pollution Abatement: A Case Study of the Nandesari Industrial Area[1]

SMITA MISRA

The rapid growth of Nandesari Industrial Estate since the seventies led to a continuous decline in water quality in the Mini-Mahi valleys and the aquifer system. This led to a number of court cases filed by the local consumer forums and State Pollution Control Board. As a result of these court cases, the Nandesari Industries Association had to undertake pollution abatement measures. These measures entail a rise in costs for the factories and the economy in general. It is essential to take these costs into account while computing the environmentally corrected national product. This chapter addresses the problem of accounting for costs of pollution abatement and examines the alternative instrument of collective action for controlling water pollution.

## 10.1 COLLECTIVE ACTION: AN ALTERNATIVE INSTRUMENT FOR INDUSTRIAL WATER POLLUTION ABATEMENT

Collective action may be seen as an alternative to fiscal instruments for controlling industrial water pollution. The need for collective action in bringing about water pollution abatement in small-scale industries is highlighted in an earlier paper (Misra and Murty, 1995a). It is an alternative institution capable of solving the 'local' problem of industrial water pollution through the 'formulation of a

[1] This chapter is a part of my doctoral dissertation to be submitted to the University of Delhi.

rule of restrained access to a common pool resource. The public good might be the resulting sustainable exploitation of the resource (Wade, 1987).

Historically, Prisoner's Dilemma, Tragedy of the Commons (Hardin, 1968) and Logic of Collective Action (Olson, 1965) can all be portrayed as unsuccessful collective action models. Their emphasis is on difficulties of voluntary collective action based on moral commitments, habits, individual benefits and the free-rider problem. However, Coase (1960) and Becker (1983) would successfully argue that voluntary collective negotiation and competition of political interest groups help in correcting market failure. Demsetz (1967) argues how 'property rights' develop to internalize the externalities when gains from internalization become larger than the costs of internalization. Runge (1981), Wade (1987) and Murty (1995) explain factors responsible in successfully bringing about collective action. One school of thought believes that the establishment of full private property rights over the commons is a necessary condition for avoiding the tragedy. Another school of thought necessitates the allocation of full authority to regulate the commons to an external agency, generally the government, to avoid the tragedy of exploitation of the commons. Given exclusive possession as one extreme on a continuum of property rights and no property rights at the other extreme, common property resource lies in between with the right to exploit a resource in common with others.

Common property and open access are often confused in the literature. Tragedy of the commons is referred to as 'open access' where a right of inclusion is granted to anyone who wants to use the resource. Under 'common property', a right of exclusion is assigned to a well-defined group and the agents are allowed to interact strategically with each other. In the case of 'open access' the agent would think that the final outcome is independent of their individual decisions whereas under 'common property' they expect that their action will induce a particular reaction from other agents, thus affecting the collective result (Baland & Plateau, 1994). Given strategic interdependent choice, institutional rules would provide perfect expectation of the other agents and assure the fulfillment of collective objectives.

Coase Theorem implies that collective action through regulation is not needed. Agents will make up for missing markets through voluntary negotiations. The Coase theorem suffers from a number of limitations restricting its practical relevance, especially for a developing economy.

1. Costs Incurred to Force Coasean Solution

*Transaction Costs*

Coase assumes costless market transactions. However, in the real world and especially in the developing economy context, the cost of altering and combining rights through the market to achieve efficient bargaining might be so great that the optimal arrangements of rights and the consequent preservation of the environment may never be reached. The State then plays an important role. It designs the initial allocation of rights, and by doing so might incur lower transactions cost than private agreements.

*Private Information*

Coase Theorem assumes complete information. When there are information asymmetries, the parties will have incentive to cheat and reveal false information thereby making bargaining inefficient. Thus it is necessary for the government to induce people to reveal the true information through an incentive scheme and to render participation to the scheme compulsory (Farrell 1987). Farrell argues that such a scheme might not be efficient but then the comparison is between the inefficient outcome of a private bargaining process and the inefficient outcome of an uninformed central authority. He shows that 'the bumbling bureaucrat outperforms both allocation of rights', an inefficient centralized decision-making process yields a better outcome than private bargaining.

2. Bargaining Power

It is not clear what role bargaining power plays in the Coasean framework. All parties concerned in the Coasean framework should be in a position to negotiate. However in the water pollution abatement case it is the future generation that bears the cost of the externality and it is not clear how its interests may be taken into account in the bargaining process.

3. Distributive Effects

Coase's Theorem bears on the issue of efficiency. The distributive effects of private bargaining are not considered. The willingness to pay might be different from willingness to accept because the former takes into account the consumer's income. Hence the assignment of property rights would affect resource use.

While the Coase solution (which might even be a Nash equilibrium) may be at one end of a spectrum with self-enforcing behaviour of the individuals, contractual arrangements with complete State regulations could be at the other end. Self-interest drives the former while an external enforcer drives the latter. The latter framework depends on the extent of distortions in the market and the failure of Coasean assumptions. In between would lie a whole string of arrangements to influence the outcome. I am considering one such arrangement. In the Nandesari context, collective action cannot be strictly defined in the sense of a fully voluntary negotiation leading to efficient abatement; rather it is a response to a judicial pronouncement on complaints filed by the affected parties.

Given the judicial pronouncement, two outcomes are possible: (1) Culprits abate voluntarily or (2) Administration and policing is required. This paper defines (1) as the (limited) collective action case.

Hence it can be said that the Coase Theorem suffers from serious practical limitations. A centralized intervention is necessary which will reduce the transaction costs of negotiation. The problem of collective action can be summed up (Ostrom, 1990) as one of finding solutions to (a) the problem of supplying a new set of institutions; (b) the problem of making credible commitments; (c) the problem of mutual monitoring.

Therefore it is important to look for the institutional set-up required for the Coasean bargaining approach that would result in a co-operative solution in practical situations. Classical Prisoner's Dilemma type of game portrays non-cooperation among agents as a dominant strategy in the case of an externality type of situation. However in an 'assurance' game in which each player co-operates when the other does, but defects when the other defects, co-operation is possible. Even the repeated Prisoner's Dilemma model would attain the same results when co-operative equilibria are sustained by the long-run interests of foresighted self-interested individuals. Also, if sufficient credible punishments are administered to outweigh the immediate benefits of the defector, the co-operative equilibrium is sustained in the repeated Prisoner's Dilemma Model. This again reflects the empirical verification in the Nandesari context where the

State Pollution Control Board's threat of closure and penalties leads to a co-operative solution among the group of polluters. Literature on collective action determines the attainment of the co-operative solution on the following points:

- the rates of time discount not being too large.
- the rewards to defection not being too large.
- the degree of trust the players have in each other.

In the real situation, there is ongoing interaction among agents in updating the rules of the game resulting in changes in the pay-off matrix. Also the role of the government, NGOs, leaders, etc. as catalysts in initiating co-operation is not brought out in the above models (Bardhan, 1993). Ostrom (1990) specifically considers 'enforcers' of co-operative arrangements, missing in the theoretical models. Field studies also show that self-interest coupled with the norm of fairness enhances co-operation. Also norms solidify easily when conflict is perceived vis-à-vis outsiders rather than intra-group conflicts (Wade, 1987). Co-operation works better in small groups with similarity of needs, clear boundaries, shared norms, patterns of reciprocity (Bardhan, 1993). Ecological stress would also determine the degree of success of co-operation.

In the following sections of this paper, some of the problems discussed above are examined in an empirical setting with the help of a case study of collective action among small-scale industries in the Nandesari Industrial Estate of Gujarat in India.

## 10.2 BENEFITS AND COSTS OF WATER POLLUTION ABATEMENT IN NANDESARI

This chapter explores collective action possibilities among small-scale industries in an industrial estate. The important 'agents' in the Nandesari political economy for water pollution abatement are (a) polluters, (b) affected parties, (c) governments at different levels and (d) NGOs. The costs and benefits to the different agents could be examined on the basis of the following 'scenarios' at Nandesari.

This study focuses on the collective action of polluters only. Collective action of other agents has been dealt with in separate studies for the same area. Collective action is the product of strategic playing of above agents in water pollution abatement involving (a) collective action by the affected parties, (b) collective action by the polluters, and (c) bargaining between the affected parties and the coalition of industries. Experiences of collective action in several industrial estates including Nandesari show a particular sequence of events pertaining to collective action by different groups. The collective action of affected parties precedes the collective action of polluters/industries and finally results in bargaining between the affected parties and the coalition of industries.

Table 10.1
Costs and Benefits without Abatement

| Agents | Costs | Benefits |
|---|---|---|
| Polluters | • Charges paid to pollution control boards for not meeting standards/threat of closure<br>• Bribes paid to local authorities | • Savings in costs incurred by not meeting standards |
| Affected Party | *Damages due to water pollution*<br>User: Water-borne diseases<br>Increases in costs of drinking water<br>Non-user: Degradation of water acquifer<br>Degradation of rivers Mahisagar, Mini<br>Degradation of Gulf of Cambay<br>Degradation of soil fertility and increase in toxicity | • Employment and income generation due to chemical industries<br>• Infrastructural development benefits due to proximity of industries |
| Government | • Cost of regulation<br>• Costs incurred in monitoring effluents. | |
| NGOs | • Costs incurred in educating people, building awareness for preservation of water quality | |

As a result of water pollution of river Mini, river Mahisagar and the aquifers, the surrounding villagers regularly complained to the local authorities, finally resulting in an enquiry of the pollution of river Mahi by the authorities in 1968. Meanwhile the Government of India started showing concern about the degradation of environmental resources, and the Water (Prevention and Control of Pollution) Act, 1974 marked the beginning of governmental regulations for the control of environmental pollution in India. Subsequently, the

Water (Prevention and Control of Pollution) Cess Act, 1977 and the Environment (Protection) Act, 1986, were also passed. The regulations have been responsible for the GIDC (Gujarat Industrial Development Corporation) constructing an effluent channel to convey the wastewater to the Gulf of Cambay in 1983. Given the nature of small-scale industries with financial, technical and space problems, the GIDC also constructed a Common Effluent Treatment Plant in 1984. The Gujarat Pollution Control Board (GPCB) also came into existence and standards were laid down for the effluents.

Table 10.2
Costs and Benefits with Abatement

| Agents | Costs | Benefits |
|---|---|---|
| *Polluters* | | |
| Without collective action of polluters | Costs of unilateral treatment by polluters | |
| With collective action by polluters | Costs of in-house, primary, and secondary treatment to meet standards | Savings in costs due to economies of scale |
| *Affected Party* | • Transaction costs of collective action and costs of organizing a club<br>• Costs of legal action | Effluent control costs ensure user benefits like<br>• savings due to prevention of water-borne diseases<br>• savings in cost in the supply of drinking water<br>• savings due to increase in productivity of soil<br>• savings due to avoidance of loss in fish productivity; and non-user benefits like<br>• existence and bequest values due to preserved estuary and soil conservation |
| *Government* | • Costs incurred in providing a catalytic role to ensure | Saving in costs of enforcement and |

(*Cont.*)

| | | |
|---|---|---|
| | collective action:<br>• Incentives in the form of financial support<br>• Providing technical know-how<br>• Costs incurred for continuous legal threats to defaulters | policing |
| NGOs | • Transaction costs incurred in filing legal cases against defaulters<br>• Costs incurred in educating people about the need for continuous water quality preservation<br>• Costs incurred in providing technical information to polluters | |

The presence of scale economies characteristic of water pollution abatement necessitated the collective action among industries instead of unilateral actions for pollution abatement (Misra and Murty, 1995a). The heterogeneity of industries in the estate called for an institutional arrangement in which there is in-house treatment, common treatment by a sub-coalition of industries and treatment by the grand coalition for water pollution abatement. An 'in-house' treatment of effluents became a necessary technological concern to enable a proper functioning of the CETP. Since there are frequent breakdowns of the CETP, the Industries Association isolated a group of highly acidic effluent industries in 1990 to form a sub-coalition and undertake primary common effluent treatment before undertaking common treatment for all industries. The industries' association also took up the operation and maintenance of the CETP from GIDC in November 1994 to ensure smooth working.

The Nandesari case illustrates how opportunism and incentives to break an agreement are avoided by co-management between the State and the community. Private ownership of the environmental good 'water quality' is not only extremely costly, but impossible in the subsistence economy of Nandesari. Direct State control is ineffective due to high information costs, lack of adequate monitoring, and short term political interests. Complete management by the local community is also not possible because local conditions might not guarantee the emergence of collective action due to changes in the rural scene, employment opportunities in the urban areas, etc. The co-management arrangement between the government and rural communities seems to work in Nandesari. Here the rural community or the user group is the primary base for resource management. The government plays a critical role due to the limitations of the local level collective action through the following:

- Processing information on 'water quality'.
- Providing technical and financial assistance in setting up the CETP, effluent channel, etc.
- Providing economic incentives through penalties and shut down threats.
- Providing a legal framework ('polluter-pays' principle) to legally enforce recognition of their identity and rights and call upon the State as the enforcer of last resort, (Wade, 1987).
- Providing a monitoring framework.

The local community (affected parties) and the polluters are also able to advantageously exploit co-management arrangements firstly, by being well informed about local 'water quality' conditions, fish loss, cattle loss, diseases, soil degradation, aquifer degradation, etc. and secondly, they can make community-specific rules taking into account fairness and equity.

The complementarity between collective action for the preservation of 'water quality', and the interests of local communities and the government is clearly seen. It also highlights the 'catalytic' role of the government instead of the 'coercive' role in bringing about water pollution abatement. The causes for the emergence of collective action by industries can be summed up as follows:

- Collective action of the affected parties.
- Catalytic role of NGOs and the government.
- Presence of a Property Rights regime which supports the 'polluter pays' principle.
- Presence of governmental regulations implemented through State Pollution Control Boards.
- Presence of scale economies characteristic of water pollution abatement.

In this way collective action involving the affected parties, the industries and the government is now seen as an institutional alternative to deal with the problem of water pollution abatement in Nandesari. This supports the 'Coase-in-politics' solution (Becker, 1983). Becker suggested that competition of political interest groups, pursuing their self-interests provides government policies that help correct market failures. Hamilton (1993) has shown that the installation of effluent treatment facilities in the USA are influenced by the presence of politically active groups in a locality. The emergence of active political groups depends upon the benefits each group gets from the bargaining (Murty, 1995a). The CETP kind of technology at Nandesari confers the benefits of savings in costs to the industries and the reduction in damages to the affected parties.

The environmental good 'water quality' associated with the industrial effluents is inextricably linked with other natural resources of the region—soil, vegetation, water resources, rivers Mini and Mahisagar, aquifers and the Gulf of Cambay. Hence water pollution abatement is a question of social relations between individuals with vested common interests. Since there are damages (user and non-user) to the affected parties (explained in the chart) they will influence the behaviour of the polluting industries. The collective action of the industries is a response to the action taken by the affected parties.

## 10.3 WHY DO INDUSTRIES CO-OPERATE?

The central question then is why should the polluters co-operate via collective action, why not defect? This could be analysed in terms of (1) economies of scale and non-separable costs through collective action; (2) mutual expectations; (3) institutional framework; (4) sustainability.

### Economies of Scale

In order to reap the benefits of economies of scale in water pollution abatement the industry's cost function has to have certain non-separabilities. It depends on the joint costs in water pollution abatement technology. Hence the individual firm seeks collective action to ensure cost minimization. The economies of scale act as externalities leading to an optimum investment in the abatement technology and the emergence of a voluntary co-operation by the firms.

### Mutual Expectations

Given the economies of scale and the institutional arrangement (in-house, primary, secondary), the propensity to co-operate depends on the expectations of the behaviour of other firms. If the firm does not expect mutual co-operation there will be a tendency to defect and Prisoner's Dilemma being enforced. However, here there are external incentives to co-operate. The State Pollution Control Board has set certain standards which have to be met. The Secondary Treatment Plant has certain standards for compliance to enable a smooth operation of the plant. Further, the cases filed by the State Pollution Control Board (forty in the last 5 years) act as external threats. The standards and the cases filed in the court ensure the joint treatment of water pollution (Sen, 1967; Runge, 1981, 1984, 1986). The Association of the Nandesari Industries is then responsible for water pollution abatement and answerable to the State Pollution Control Board for complying with the standards. Thus the Association promotes the 'critical mass'[2] which assumes the responsibility to preserve 'water quality'. Cooperation is thus consistent with self-interested behaviour and the Prisoner's Dilemma framework is resolved with repeated plays and by acquiring information about others' behaviour (Hardin, 1982). Also, the principle of reciprocity (Sugden, 1984) would operate to avoid free riding when others are co-operating. Thus commitment to a rule of behaviour depending on mutual expectations leads to a 'critical mass' resolving the abuse of the public good, 'water quality'.

## Institutional Framework

The institutional arrangement ensures a fair share of each industry in the collective participation of water pollution abatement. The total gains from the CETP are shared on a mutually agreeable 'fair' basis. The cost effectiveness is on the basis of pollutants. Thus primary treatment becomes necessary for highly acidic industries. The group size and the heterogeneity problem is also solved by the institutional arrangements of a Common Primary and a Common Secondary Treatment Plant. Having two homogeneous groups with similar effluents and thus similar interests resolves the problem of non-cooperation among the heterogeneous groups. It reinforces reciprocal behaviour which is Pareto optimal and adds to the assistance for motivating a joint treatment of water pollution. This chapter empirically verifies the Coase-in-Politics solution given by Becker. It

[2] 'Critical mass' refers to the agents actually responsible for control of water pollution.

supports co-operative behaviour revealed in a number of game theoretic frameworks (Bardhan, 1993). Findings of this paper suggest that 'pre-game' communication can induce co-operative behaviour. Collective action of the affected parties and the intervention of the State helped in inducing co-operative behaviour of the polluters. Credible threats by the Pollution Control Board and retaliatory action by the affected parties enforce the collective behaviour of the polluters. The required transformation in the pay-off matrices was accomplished by (a) creating selective incentives and (b) creating appropriate social norms and standards. Finally, assurance game situation leads to the formation of a sub-coalition of industries for primary treatment in addition to a grand coalition for secondary treatment.

### Sustainability

The question of sustainability in dynamic situations arises with the possibility of multiple equilibria. The environment conditions, group membership and objectives are likely to change over time. Rural conditions change with alternative job opportunities, better markets and better income. Also over time selective incentives may be especially useful. The State and the rules it establishes and enforces could strengthen and increase the efficiency of collective action. The relationship between the State, the market and the local institution of collective action could be complementary. Appropriate linkages should be developed for the preservation of 'water quality'.

## 10.4 THE INSTITUTIONAL ARRANGEMENT SUPPORTING COLLECTIVE ACTION

The Nandesari Industrial Estate consists mainly of two kinds of industries—chemicals and pharmaceuticals. Each industry has an 'in-house' treatment plant for eating effluents before sending them for common treatment.

Twenty-six industries set up a Nandesari Industries Primary Treatment Plant in co-ordination with GIDC and GPCB. It was set up at a cost of Rs 1.5 crores in 1992. The primary treated effluent is then conveyed to the Secondary Treatment Plant at Nandesari. The wastewater influent to the Primary Treatment Plant from the twenty-six industries hardly gets any in-house treatment. Most of these industries produce Vinylsulphone, H-acid, Reactive dyes, J-acids, Sulphuric acid, S-sulpho anthranic acid, etc. which have been banned abroad. These 'in-house' effluent treatment plants consist of 'collection tanks' primarily to remove some suspended solids. They rely on the primary treatment plant for their wastewater treatment. The pH of the sample wastewater of these industries ranges between 0.25–0.35. It is supposed to be the strongest acidic waste and for this reason a special primary treatment plant is needed to ensure that effluent parameters meet secondary treatment plant requirements.

The Secondary Treatment Plant was constructed in 1984 by GIDC but did not work on a continuous basis due to various operational reasons. However the Nandesari Industries Association took over the plant in November 1994 and it commenced operations since March 1995. It treats about 2,400 kl of wastewater per day, generated from all the 250 units at Nandesari Industrial Estate. The treated effluent is discharged into the Effluent Channel Project (ECP) which finally carries it over to a distance of 56 km before discharging it into the Gulf of Cambay.

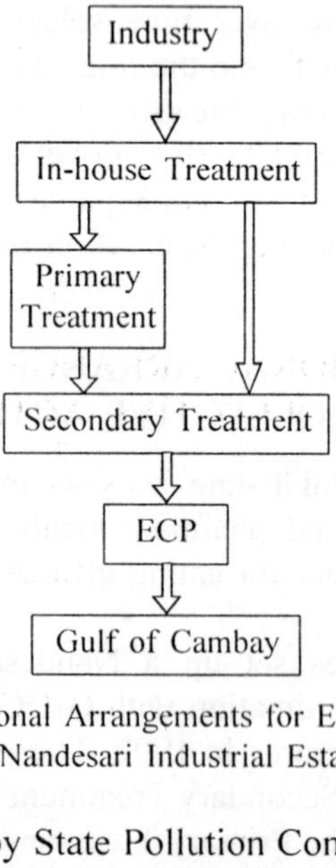


Fig. 10.1 Institutional Arrangements for Effluent Treatment in Nandesari Industrial Estate

The standards set by State Pollution Control Board (SPCB) (See Table 10.3) ensures the collective action in treating water pollution via an institutional arrangement at Nandesari. This answers the question of mutual monitoring for successful collective action (Ostrom, 1990). The free rider problem gets automatically resolved. Since the industry has to meet the SPCB standards, it is in the interests of the association to make sure the CETP works. Thus the standards of the CETP have to be met by in-house and primary treatment, otherwise the CETP will break down. Even if an industry is tempted to free ride and dump its wastes into the river Mini, the effluent starts accumulating in the open and the association would have to take action to avoid confrontation with the GPCB.

Table 10.3
Effluent Standards as on March 1995
(except pH; All values are in mg/litre)

| Para-meters | In-House After Treatment | Primary After Treatment | Secondary | | ECP | SPCB |
|---|---|---|---|---|---|---|
| | | | Before | After | | |
| BOD | 2730–6336 | 2000–3000 | 1750–2100 | 123–220 | 100 | 30 |
| COD | 10894–21120 | 8000–9000 | 2800–4500 | 240–450 | 250 | 100 |
| SS | 3740–11721 | 600–800 | 400–750 | 112–210 | 100 | 100 |
| PH | 0.25–0.75 | 6.6–8.5 | 6.5–8 | 7–7.5 | 6.5–8.5 | 6.5–8.5 |

There have been about forty cases filed against the industries by the GPCB since 1986. This includes the case against the 26 highly acidic industries. The President of the Association has to guarantee the proper working of the in-house, Primary, and Secondary Effluent Treatment Plants. Ultimately the association is responsible for the effluent control at Nandesari. This ensures that everyone co-operates and the free rider problem is resolved.

The association has agreed on the following cost-sharing arrangements:

• In the case of the Primary Treatment Plant, the annualized fixed costs are shared on the basis of net profits as follows:

Table 10.4
Share of Fixed Costs of Primary Treatment Plant

| Net Profits | Percentage of Annualized Fixed Costs to be Shared |
|---|---|
| 50 lakhs and above | 80 per cent |
| 20–50 lakhs | 10 per cent |
| 10–20 lakhs | 7.5 per cent |
| 0–10 lakhs | 2.5 per cent |

• The variable costs are shared on the basis of Rs 0.50 per kilolitre of water consumed. This ensures cost sharing on a mutually agreeable basis.

• In the case of the Secondary Treatment Plant, the fixed costs are shared on the basis of Rs 3 per kilolitre of water consumed. The variable costs are shared on the basis of Rs 1.30 per kilolitre of water consumed. The annualized fixed costs are considered for estimating effluent control costs. The Secondary Treatment Plant ensures cost sharing on a proportional basis.

The cost sharing arrangement solves the problem of making credible commitments in the problem of collective action to obtain long-term collective benefits. The institutional and the cost-sharing arrangements ensure sustainability of the collective action of the polluters.

## 10.5 COSTS AND BENEFITS: INSTRUMENTS AND TECHNIQUES

It has always concerned economists that while the production of goods and services might continue to increase, the 'quality of life' might not necessarily improve. While material goods increase, the quality of physical environment like water might deteriorate due to water pollution caused by industrial activities. The question then arises is on what basis should decisions be made regarding choices entailing action on the environment. Given the public good characteristic of the environmental good, 'water quality', and the free rider problem associated with it, the market fails to generate a true price of the good. One cannot arrive at a value of this amenity directly and one of the objectives of this paper is to explore the problems pertaining to the estimation of this value.

Economists have always suggested that firms discharging effluents into the environment should somehow be made to pay a price for such discharges depending on the amount of environmental damages caused. Hence the environmental damage should be estimated. The environmental damages in the case of water pollution could be estimated either by a 'replacement cost approach' or by a 'preventive expenditure approach'. The rationale for the 'replacement cost approach' is to consider the costs incurred to replace productive assets which are damaged. The 'preventive expenditure approach' considers the defensive expenditures incurred by the consumers to reflect the minimum value of benefits generated. While the data for the former approach would be impossible to generate,[3] the latter approach is unlikely in the subsistence economy of Nandesari.

An alternative estimate of costs of pollution gradually gaining ground in the literature is the estimation of 'effluent control costs' to reflect the amount of damages incurred. Needless to say, the effluent control cost might not reflect the actual damages incurred. However it would provide a good indicator of costs of preservation which can be compared with an estimate of 'benefits' from preservation. The problem of estimation of the benefits of preservation of 'quality of water' in Nandesari Industrial Area has been dealt with in the earlier chapter. This paper is concerned with the estimation of costs of preservation in terms of the 'effluent control costs'. In the context of this paper, it is relevant to examine the question of who pays the costs associated with pollution. This could be answered only if one distinguishes between the liabilities to different agents arising out of 'pollution' according to specified 'pollution rights'.

Given the 'marginal control' costs as a decreasing function of the effluent load (Misra and Murty, 1995a), and 'marginal environmental damages' as an increasing function of the effluent load, Fig. 10.2 explains the different concepts of pollution. Pollution occurs whenever there is a perceived loss in environmental value due to effluents discharged. Given the control costs and the environmental damages associated with pollution, the 'optimal effluent load' is one which minimizes the sum of control costs and environmental damages. The 'optimal pollution' is the environmental effect of optimal effluent load. Thus the optimal pollution would also define the 'acceptable pollution' CQ.

OC is the maximum effluent load that can be absorbed without causing perceived environmental damages. If the polluter has to pay for all the perceived damages then the first definition of pollution holds, that is pollution is zero in the range OC and then positive. The polluter then compensates for 'acceptable' pollution and 'unacceptable' pollution. The 'polluter pays' principle (PPP) means that the

[3] The direct replacement cost approach considers the costs which will be incurred to replace the damaged productive assets. In the case of Nandesari, the damages are related to the water quality of rivers Mini and Mahi. It is impossible to assess the economic cost of replacing the degraded water quality of the rivers using this method.

polluter should bear the expenses of carrying out the (pollution prevention and control) measures decided by public authorities to ensure that the environment is in an acceptable state (OECD, 1972). Residual pollution compensation is not obligatory (OECD, 1975). Thus the PPP requires polluters to pay for controlling effluent to the optimal load, but not for environmental damage caused by the optimal effluent load. Thus the polluter enjoys *de facto* 'pollution rights' CQ in Fig. 10.2, and in this way pollution is defined by definition 2. This gives the 'control costs' EQS as implied by PPP shown as area QES. A third definition of pollution would be to the polluter not enjoying any right and thus has to compensate for the entire damages OS. This includes OC where the effluent contains potentially damaging contaminants even though no detectable environmental damages can be perceived.

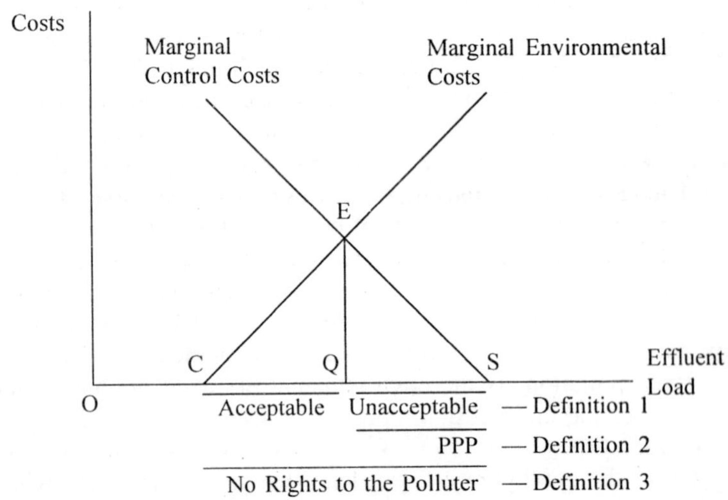


Fig. 10.2 Different Concepts of Pollution

We can thus distinguish between:

1. Control costs (implied by PPP)

2. Prevention costs (standards leading to an effluent load which is much higher than the optimal effluent load)

3. Damage costs (perceived damages to be borne by polluters)

4. Clean-up costs (complete absence of pollution rights and the entire costs OS to be borne by polluters).

How much of the pollution cost is borne by the polluter therefore depends on the 'pollution rights' enjoyed by the polluter:

1. If the polluter has the right to discharge any effluent, then at least all pollution control costs should be borne by the government.
2. If the polluter has the right to discharge optimal effluent OQ, then the control costs EQS should only be borne by the polluter.
3. If the polluter has no 'pollution rights', he should pay both the control costs EQS and compensation for pollution damages CEQ.

In the case of Nandesari Industrial Estate, the PPP holds and the polluter is required to meet the effluent control costs according to Pollution Control Board Standards. The effluent control costs would be welfare improving as long as this pollution can be reduced for a cost which is less than the benefits that such a reduction yields in terms of reduced damages. If all Pareto relevant externalities are eliminated, the gains from exchange (from the model of collective action) will be exhausted, and Pareto equilibrium will be achieved for a given distribution of income.

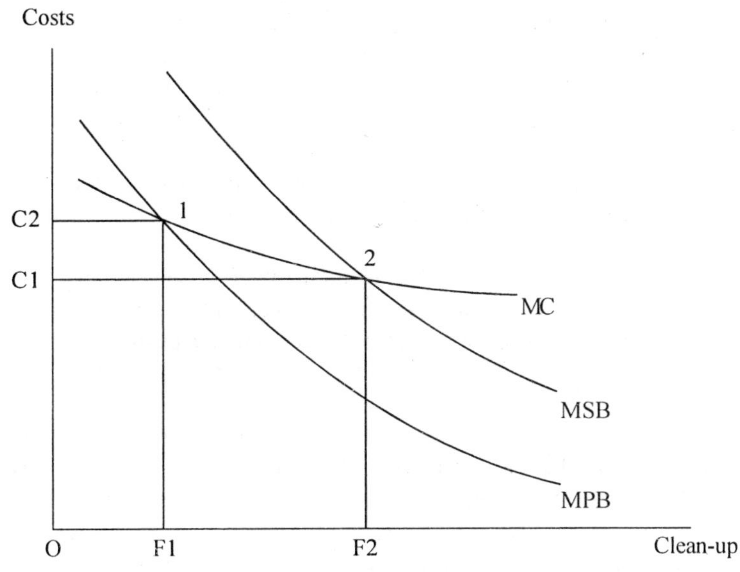


Fig. 10.3 Economies of Scale, Property Rights and Pollution Abatement

Taking into account only private costs and benefits and economies of scale in abatement technology for small-scale industries the optimal level of pollution clean-up is $F_1$ where the marginal costs (MC) are equal to the private marginal benefits (MPB). However, if we add the marginal social benefits to the private benefits, the optimal pollution clean-up increases to $F_2$. Thus clean-up could increase, given the shift of the marginal benefits curve. This would depend on: (1) property rights arrangement (2) informational asymmetries (3) transaction costs (4) policy instruments to influence behaviour.

If property rights are well-established then negotiations among the polluters and the affected parties should internalize costs and Pareto relevant externalities removed (Coase, 1960). However, in developing economies like India and in the context of Nandesari, the 'legal property rights' might be very different from the economic property rights. The latter would reflect reality which might not be reflected in 'formal property rights'. Thus the movement from 1 to 2 for Nandesari would depend on establishing legal property rights. A cost effective strategy or policy instrument to influence behaviour is then required. The role of the government at various levels and of NGOs is critical here. They would be responsible for providing adequate information as a necessary prerequisite for the effective operation of the markets and a movement from (1) to (2). The polluters have to foresee the 'economies of scale' and the 'threat from affected parties'. The information provided to the affected parties will help in providing threats. This will ultimately lead to a facilitation of negotiation and collective bargaining between the two with the attainment of the efficient level of clean-up at (2). There is a deviation from Coase's (1960) voluntary negotiation leading to the correction of market failure. Given the developing economy context and market distortions, negotiations between the polluters and the affected parties are facilitated with the catalytic role of the government/NGOs.

Given the economies of scale in water pollution abatement technology and hence the inapplicability of Pigouvian taxes (Misra and Murty, 1995a), collective action as a non-fiscal instrument could efficiently correct market failure. If the effluent control costs are Rs 1 lakh per annum for an industrial estate, and damages are worth Rs 2 lakhs, an opportunity exists for mutually beneficial trade. If the polluter has the right to pollute, he will accept anything in excess of Rs 1 lakh to clean up. The affected parties will offer anything less than Rs 2 lakhs to clean up. The actual bargain would depend on their relative strengths and skills for bargaining. If the affected parties have to clean water, the polluter will offer anything upto Rs 1 lakh. However, this will be unacceptable to the affected parties, and the polluter will have to clean up.

The Coasean framework as described above will result in efficient bargaining irrespective of who holds the property rights. However, given the problems of a developing economy in Nandesari context, a framework facilitating negotiation is a necessary condition for the collective bargaining to materialize. The role of the government/ NGOs then is crucial to provide full information to each agent about the costs and benefits. The cost of negotiation then becomes important in facilitating the framework. If the transaction costs or the costs of negotiation exceed Rs 1 lakh, the opportunity for a Pareto improvement will not be realized. Here the Coasean solution applicable to the small numbers case is sensible because the transaction costs would have to be low to enable collective bargaining.

The CAC instruments have a role in the Indian context, precisely because the incentives are ineffective and the assignment of property rights and negotiated settlements are insufficient. The standards set by the GPCB in the Nandesari context provide the necessary support needed to facilitate an efficient bargaining between the polluters and the affected parties. The CAC instruments pose external costs and the polluter would rather abate than incur these costs.

## Effluent Control Costs

The annualized fixed cost is computed on the basis of a generalized 'rate of return' and depreciation from Annual Survey of Industries data. For chemical industries the 'gross rate of return' is 31 per cent while for drugs and pharmaceutical industries it is 35 per cent. On the basis of primary data collected for a sample of 90 industries out of 250 industries at Nandesari Industrial Estate, detailed information was obtained about the cost of in-house treatment: the year of construction, the area used for the construction of the in-house treatment plant, the number of labourers employed, wages paid per annum, the fuel and power costs, the costs of material inputs used and the volume of water treated. The details of the sharing arrangements of fixed costs and variable costs for the Primary Treatment Plant and the Secondary Treatment Plant were also obtained.

The following effluent control costs per industry and for the entire industrial estate have been estimated.

Table 10.5
Annual Effluent Control Costs by Industry Groups
(Rs lakhs per annum)

| Costs | Per industry | 26 Industries | 224 Industries | 250 Industries |
|---|---|---|---|---|
| P.I.C. in-house | 1.45 | 37.7 | 324.8 | |
| P.I.C. primary | 3.56 | 92.56 | | |
| P.I.C. secondary | 0.13 | 3.38 | 29.12 | |
| P.I.C. Total | | 133.64 | 353.92 | 487.56 |
| Total Cost in-house | 2.41 | 62.66 | 539.84 | |
| Total Cost primary | 8.08 | 210.08 | | |
| Total Cost secondary | 1.91 | 49.66 | 427.84 | |
| Total Cost for Industrial Estate* | | 322.4 | 967.68 | 1,290.08 |

*Note:* PIC = Primary Input Costs.

While the total cost for effluent control for the entire industrial estate is Rs 1,290.08 lakhs per annum, the value-added foregone in terms of the primary input costs is Rs 487.56 lakhs per annum for the industrial estate. However, if the benefits[4] against these costs are estimated, the net benefits could be positive.

## Regression Results

A double log cost function has been used to estimate $C = e^{\alpha} V^{B}$ where C = cost; V = volume of water.

The regression results support the central argument of the supply, commitment and monitoring of collective action. The $\beta$ coefficients reveal economies of scale for the in-house and the secondary arrangements. The economies of scale provide a technological incentive to have the CETP for an industrial estate. The primary treatment reflects the increasing cost technology. This supports the decision of the association in isolating highly acidic industries to form a sub-coalition for primary treatment.

[4] Ch. 9 provides estimates of these benefits.

Table 10.6
Regression Results of Pollution Abatement Costs for the Industrial Estate

| | α (t-value) | ß (t-value) | $R^2$ | n |
|---|---|---|---|---|
| V.A. Loss in-House | –3.06 (–5.89) | 0.42 (6.07) | 0.35 | 69 |
| V.A. Loss primary | –4.47 (–3.54) | 0.52 (3.29) | 0.35 | 22 |
| V.A. Loss secondary | –6.32 (–17.83) | 0.53 (10.71) | 0.56 | 89 |
| V.A. Loss total | –5.34 (–10.17) | 0.71 (9.75) | 0.52 | 89 |
| In-house Cost | –3.44 (–6.96) | 0.56 (8.62) | 0.53 | 69 |
| Total Primary Cost | –6.88 (–9.24) | 1.04 (11.03) | 0.86 | 22 |
| Total Secondary Cost | –5.87 (–10.55) | 0.86 (11.03) | 0.58 | 89 |

### Extension of National Income Accounts

The total value-added loss due to water pollution abatement by the industrial estate is considered as the cost to the economy in the Conventional System of National Income Accounts (CSNA). The opportunity cost of pollution abatement in an industry is the value-added loss due to the use of capital and labour in the abatement technology rather than in the main plant. Hence the total value-added losses of Rs 487.56 lakhs for the industrial estate reflects the opportunity cost of water pollution abatement. It is the primary input cost of the abatement technology. However in the Extended System of National Income Accounts (ESNA), the preservation benefits due to these losses are accounted for. There can be net benefits from the pollution abatement.

## 10.6 CONCLUSION

The central question is how a group of polluters can organize themselves to obtain joint benefits, when all face temptations to free ride or act opportunistically. The collective action of the polluters is seen as a response to the collective action of the affected parties, the government and NGOs. Hence an alternative institution emerges to solve the problem of exploitation of a common resource 'water quality'. The paper explains how the new institution emerges, how commitments are obtained and how monitoring leads to continuous joint benefits. Bargaining between the polluters and the affected parties ultimately leads to the establishment of property rights and the benefits of preservation are conferred on the affected party. This chapter illustrates how collective action of polluters is brought about at Nandesari. First, water pollution abatement at Nandesari is brought about through collective action of the polluters where pollution abatement becomes the responsibility of the Association of Industries. Second, technological concerns best explain the individual propensity to co-operate and provide the necessary condition for collective action to emerge. Third, the mutually agreeable credible commitments and institutional arrangements provide sufficient conditions for the emergence of collective action.

Given the PPP and the CAC types of techniques used for water pollution abatement in India, this chapter attempts an economic valuation of effluent control costs in meeting standards for Nandesari Industrial Estate in Gujarat. The 'effluent control costs' are no indication of the benefits of preservation. If we compare these costs with the benefits estimated in Ch. 9, these can be net social benefits from the control of water pollution in Nandesari industrial area.

# 11. Environmental Regulation in India

## 11.1 SOME CONCLUSIONS

Chapters 1 and 2 show that while policy regimes currently in place in developing and developed countries to control and regulate industrial pollution include both command-and-control (or regulatory) and economic instruments, the evidence on impact is rather mixed. While the case of Japan illustrates the successful use of regulatory instruments, at least to combat initially high levels of pollution, the Indian case offers a contrast where nearly two and a half decades of anti-pollution legislation has not arrested the increasing trend of environmental pollution. The review of environmental policy in several developed and developing countries show that the use of economic instruments is yet to make a significant dent on environmental pollution with the existing instruments unable to pose a significant disincentive to polluting industries.

There is, at least in the case of India, a strong need to augment the existing portfolio of policy instruments to include economic or market-based instruments and other institutional methods to have a multi-layered and multi-pronged attack on the problem of environmental pollution by industrial activities.

Chapter 3 illustrates the method of generating economic instruments to tackle pollution, by providing details about the taxes-and-standards approach to industrial water pollution abatement. The call is to begin generating a comprehensive database as soon as possible, in order to have the methodological raw material to begin installing a regime of economic instruments. The chapter shows that although problems exist in the specification of these instruments, these are not insurmountable.

Chapters 4 to 7 provide arguments in favour of using Coasean bargaining for industrial water pollution abatement and present some empirical evidence in support, from recent developments in water pollution control in India. The research output reported in all these chapters supports the view that collective action of all the relevant agents can be an alternative institution to governmental instruments to control an environmental externality.

Chapter 8 provides an estimate of environmentally sustainable income or income generated after accounting for primary and intermediary input cost of water pollution abatement by the Indian industry. Chapter 9 provides estimates of benefits from water pollution abatement in an industrial estate to rural and urban dwellers. This chapter demonstrates that contingent valuation methods can be successfully used in the developing countries in eliciting the values rural and urban populations place on environmental quality. Chapter 10 presents a detailed case study of an industrial estate for estimating the cost to factories and the cost to the economy of water pollution abatement. The research output reported in chapters 8 to 10 underscores the importance of cost benefit analysis of water pollution abatement in providing the inputs for estimating the environmentally corrected GDP.

## 11.2 FUTURE RESEARCH PRIORITIES

There is an urgent need for studies about the desirability of emission taxes or emission permits to deal with urban air pollution and industrial air pollution in India. Given that urban air pollution in India is mainly vehicular pollution and it is difficult to tax emissions, it is important to examine the desirability of having a mix of instruments consisting of fuel taxes and command and control measures to deal with this problem. In the case of industrial air pollution, it may be useful to look into the case of using emission permits to control air pollution within an airshed.

For using the taxes-standards method for industrial water pollution abatement in India, the cost function for water pollution abatement is estimated in this book using data for a sample of industries drawn from seventeen highly water-polluting industries. Since the pollution tax instrument has not yet been used in India, the estimated cost function corresponds to the pollution abatement technologies used by the industries under the current regime of command-and-control measures. Once a pollution tax policy regime is introduced, industries will have incentives to use cost-minimizing abatement technologies and therefore the cost function estimated for the post tax scenario is likely to be different from the cost function estimated for the pre-tax command-and-control scenario. Therefore even after introducing the pollution tax, there is a need for estimating the abatement cost function on a regular basis, say at an interval of two to five years, so that technological progress in pollution abatement can be taken into account while fixing the rate of pollution tax.

There is a need for research on an integrated approach involving industries, households and government to improve the quality of water in Indian rivers. It is useful to examine the case for having a mixture of economic instruments and command-and-control measures as part of the major river conservation programmes like the Ganga Action Plan. A combination of marketable permits and command-and-control methods can offer a very flexible set of instruments to deal with household and industrial effluents in different reaches of a river with varying ambient water quality.

Given that collective action involving local people, industries and government in industrial pollution abatement has been effective in some industrial estates having small-scale enterprises, it is important to examine the possibility of replicating the same in other industrial areas in India. Given that formal regulation is weak due to various reasons, and the book finds that compliance of industries with pollution abatement regulations is more in locations where local communities are politically and economically strong, it is important to look for ways and means of strengthening and increasing the awareness levels of local communities.

## 11.3 INDIA: RECOMMENDATIONS FOR POLICY REFORM

It is time India switched to using economic instruments like pollution taxes and marketable pollution permits to deal with its environmental problems. For practical purposes, however, it may be more effective to use a mix of market-based or economic instruments and command-and-control measures.[1] While the comprehensive set of environmental legislation that India has tabled so far will provide the necessary legal institutions for successfully implementing such

[1] See also Murty (1997) ESCAP.

instruments, the use of economic instruments will remove the production inefficiencies associated with the command-and-control measures used so far, and facilitate the choice of least-cost pollution abatement technologies by polluting industries.

Empirical work reported here has illustrated that a pollution tax based on the taxes-standards method for industrial water pollution abatement can result in effective control of pollution by big factories. However, this illustration also heightened the need for a revision of the current concentration-based system of pollution standards. The need is for load-based standards which are detailed for regions (i.e., environmental sinks) and industries (i.e., pollutants). Given the federal structure of India, the existing institutions of the State Pollution Control Boards may have to be entrusted with the responsibility of design and implementation of (region-specific) pollution standards and taxes, with research support from professionals either within or outside the government. Overall coordination will have to be done by the Central Board.

While the problem of big industries can be tackled by such a tax system, the large number of industrial estates consisting of small-scale industries cause environmental pollution problems which cannot be so tackled in a similar manner because of the presence of scale economies in pollution abatement, especially in water pollution abatement. In these cases, incentives will have to be provided for these industries to organize themselves sufficiently to construct a CETP to reap the benefits of scale economies. Although several such CETPs have been established in different parts of India, with subsidized governmental support, the nature of government support has to be re-examined and revised for greater effectiveness.

In the case of small industries, as well as in the general case of pollution by larger industries, there is a pressing need for legal and financial support, either from governmental or non-governmental organizations, to promote collective action which is one of the most effective institutional means of combating industrial pollution. In the political economy of environmental pollution abatement, collective action of not only those affected by pollution, but also those generating pollution (such as small-scale industries) are a vital key to environmental protection, especially in the case where the government itself is either apathetic, non-benevolent or corrupt.

In addition, serious efforts have to be made to promote the environmental education of the general public so that people are better able to judge and influence the choices and locations of environmentally harmful projects. While the recent and spectacular growth of public litigation cases related to environmental problems in the country and the success of some of the cases are evidence of increasing environmental awareness of the general public, and while recent legislation of the Government of India empowering local people to claim damages from environmental pollution and also decide about the location of environmentally harmful projects are steps in the right direction in this context, many such steps will have to be taken in the near future and at a rapid pace, if India is to overtake and successfully combat the spreading menace of environmental pollution. The people, after all, are the government, and the country and its resources are theirs to preserve into the future.

# References

ADAR, Z. and J. M. GRIFFIN (1976): 'Uncertainty and the Choice of Pollution Control Equipment', *Journal of Environmental Economics and Management*, vol. III, Oct.

ARROW, K. J. (1970): 'The Organization of Economic Activity: Issues Pertinent to the Choice of Market Versus Non-Market Allocation', *in* M. J. Markham and R. H. Haveman (eds), *Public Expenditure and Policy Analysis*, Chicago.

ATKINSON, A. B. and J. E. STIGLITZ (1980): *Lectures on Public Economics*, MacGraw Hill, New York.

BAIL, K. and J. F. SHOGREN (1994): 'Reimbursements for Citizen Suits', *Journal of Environmental Economics and Management*, vol. 21.

BALAND, J. M. and J. P. PLATEAU (1994): *Halting Degradation of Natural Resources: Is there a Role for Rural Communities*, Oxford University Press, Oxford.

BARDHAN, P. (1993): 'Analytics of the Institutions of Informal Co-operation in Rural Development', *World Development*, vol. 21, no. 4, pp 633–9.

BARTELMUS, P., E. LUTZ and S. SCHWEINFEST (1992): *Integrated Environmental and Economic Accounting–A Case Study of Papua New Guinea*, CIDE Workshop on Environmental Economics and Natural Resource Management in Developing Countries, World Bank, Washington D. C. 22–4 Jan.

BATOR, F. (1958): 'The Anatomy of Market Failure', *Quarterly Journal of Economics*, vol. 72, pp 351–95.

BATSTONE, R. et al. (eds) (1989): 'The Safe Disposal of Hazardous Wastes', Vol. II, World Bank Technical Paper No. 93.

BAUMOL, W. J. (1972): 'On Taxation and Control of Externalities', *American Economic Review*, vol. 42, no. 3, pp 307–22.

BAUMOL, W. J. and D. F. BRADFORD (1972): 'Detrimental Externalities and Non-convexity of Production Sets', *Economica*, vol. 39.

BAUMOL, W. J. and W. E. OATES (1988): *The Theory of Environmental Policy*, 2nd edn, Cambridge University Press, Cambridge.

BECKER, G. S. (1983): 'A Theory of Competition Among Pressure Groups for Political Influence', *Quarterly Journal of Economics*, vol. 98, pp 371–400.

BISHOP, R. and T. HEBERLEIN (1979): 'Measuring Values of Extra Market Goods: Are Indirect Measures Biased?' *American Journal of Agricultural Economics*, vol. 61, no. 5, pp 926–30.

BOEHMER CHRISTIANSEN, S. A. (1990): 'Energy Policy and Public Opinion: Manipulation of Environmental Threats by Vested Interests in the UK and West Germany', *Energy Policy*, Nov.

BOERO, G. and A. SILBERSTON (1995): *Environmental Economics*, Confederation of European Economic Associations Conference Volume, St. Martin's Press, New York.

BOHM, P. (1997): 'Efficiency Issues and Montreal Protocol on CFCs', *in* P. Dasgupta and K. G. Maler (eds), *The Environment and Emerging Development Issues,* Cambridge University Press, Cambridge.

BOOKSHIRE, D. S., A. RANDALL and J. R. STOLL (1980): 'Valuing Increments and Decrements in Natural Resource Service Flows', *American Journal of Agricultural Economics*, vol. 62, No. 3, pp 478–88.

BOVENBERG, A. L. and R. A. DE MOOIJ (1995): 'Do Environment Taxes Yield a Double Dividend', *in* G. Boera and A. Silberston (eds), *Environmental Economics*, St. Martin's Press, New York.

BRISCOE, J. et al. (1990): 'Towards Equitable and Sustainable Rural Water Supplies: A Contingent Valuation Study in Brazil', *The World Bank Economic Review,* vol. 4, no. 2, pp 115–34.

BROMLEY, D. W. and D. P. CHAPAGAIN (1984): 'The Village against the Centre: Resource Depletion in South Asia', *American Journal of Agricultural Economics*, vol. 66.

BROMLEY, D. W. (1992): *Environment and Economy: Property Rights and Public Policy*, Blackwell, Oxford, UK and Cambridge, USA.

BURGESS, R. and N. STERN (1992): 'Taxation and Development', Development Economics Research Programme, No. 42, London School of Economics, London.

BUSH, W. C. and L. S. MAYER (1974): 'Some Implication of Anarchy for the Distribution of Property', *Journal of Economic Theory*, vol. 8.

CAMERON, T. A. (1988): 'A New Paradigm for Valuing Non-Market Goods Using Referendum Data: Maximum Likelihood Estimation by Censored Logistic Regression', *Journal of Environmental Economics and Management,* vol. 15, no. 3, pp 355–79.

CPCB (1989): *Pollution Control Acts and Rules with Amendments*, Central Pollution Control Board, Government of India, Delhi.

——— (1995): *Standards for Pollutants,* Central Pollution Control Board, Government of India, Delhi.

CHIANG, P. C. (1994): 'Balancing Industrial Development and Environmental Protection: the Use of Command-and-Control (CAC) and Economic Instruments for Environmental Protection–The Taiwan (ROC) Experience', Paper prepared for the IGFA Mini-Workshop on the Newly Industrialized Countries and Global Change Research, Taipei, Taiwan (ROC), 24 Jan.

CHOPRA, K., G. K. KADEKODI and M. N. MURTY (1990): *Participatory Development: People and Common Property Resources*, Studies in

Economic Development and Planning, No. 52, Institute of Economic Growth, Sage Publications, Delhi.

COASE, R. H. (1960): 'The Problem of Social Cost', *The Journal of Law and Economics*, vol. III, pp 1–44.

———— (1988): *The Firm, the Market and the Law*, Chicago University Press, Chicago.

Commission on Environmental Accounting, Ministry of Finance, Stockholm (1991): *Taking Nature into Account: Proposed Scheme of Resource and Environmental Accounting.*

CORNES, R. and T. SANDLER (1986): *The Theory of Externalities, Public Goods, and Club Goods,* Cambridge University Press, Cambridge.

CROPPER, M. L. and W. E. OATES (1992): 'Environmental Economics: A Survey', *Journal of Economic Literature,* vol. 30.

DALES, J. H. (1968): *Pollution, Property and Prices,* University of Toronto Press, Toronto.

DASGUPTA, A. K. and M. N. MURTY (1985): 'Economic Evaluation of Water Pollution Abatement: A Case Study of Paper and Pulp Industry in India', *Indian Economic Review*, vol. 20, no. 2, pp 231–67.

DASGUPTA, P. and G. HEAL (1979): *Economic Theory and Exhaustible Resources,* Cambridge University Press, Cambridge.

DASGUPTA, P. (1982): *Control of Resources*, Basil Blackwell, Oxford.

DASGUPTA, P., B. KRISTROM and K. G. MALER (1994): 'Current Issues in Resource Accounting', *in* P. Dasgupta and K. G. Maler (eds), *Current Issues in Environmental Economics.*

DASGUPTA, P. and K. G. MALER (1997): *The Environment and Emerging Development Issues*, vols I & II, Manchester University Press, Manchester, 1997.

DEMSETZ, H. (1967): 'Toward a Theory of Property Rights', *American Economic Review*, vol. 57.

Department of Science and Industrial Research, Government of India (1993): *Technology in Indian Water and Effluent Treatment Plant Industry.*

DIAMOND, P. A. and J. HAUSMAN (1994): 'Contingent Valuation: Is Some Number Better than No Number?', *Journal of Economic Perspectives*, vol. 8, no. 4, pp 45–65.

DIAMOND, P. A. and J. A. MIRRLEES (1971): 'Optimal Taxation and Public Production, I and II', *American Economic Review*, vol. 61, pp 8–27 and 261–78.

DORNBUSCH, R. and J. POTERBA (eds) (1991): *Economic Policy Response to Global Warming,* forthcoming.

*Economic Journal* (July 1991): vol. 101, No. 407, 'Policy Forum: Economic Aspects of Global Warming', pp 902–48.

ELLIOTT, R. and J. SEAGRAVES (1972): 'The Effects of Sewer Surcharges on the Level of Industrial Wastes and the Use of Water by Industry', Water Resources Research Institute, Report No. 70, Aug.

EPAT (Environmental and Natural Resources Policy and Training Project, Technical Assistance Component, Winrock International Environmental Alliance for the US Agency for International Development) (1993): 'Use of Economic Instruments for Environmental Protection in Developing Countries', in *Economic Instruments for Environmental Management in Developing Countries,* Organization for Economic Cooperation and Development (OECD), Paris.

FARRELL, J. (1987): 'Information and Coase Theorem', *Journal of Economic Perspectives*, vol. 1, no. 2, pp 13–29.

FAULHABER, G. R. (1975): 'Cross-Subsidisation: Pricing in Public Enterprises', *American Economic, Review*, vol. 165, pp 966–77.

FISHELSON, G. (1976): 'Emission Control Policies under Uncertainty', *Journal of Environmental Economics and Management*, vol. III.

FISHER, C. (1956): 'Cincinnati Industry Reduces Sewer-Surcharges', *Sewage and Industrial Wastes,* 28 Sept., pp 1186–7.

FOSTER, J. F. (1981): 'The Fundamental Principles of Economics', *Journal of Economic Issues*, 15, Dec.

FRAAS, A. G. and V. G. MUNLEY (1984): 'Municipal Wastewater Treatment Cost', *Journal of Environmental Economics and Management,* vol. 11, pp 28–38.

FREEMAN III, A. M. (1993): *The Measurement of Environmental and Resource Values: Theory and Methods,* Resources for the Future, Washington, D. C.

GOLDAR, B. N. (1996): 'The Role of State and Market: The Case of India', mime., Institute of Economic Growth, Delhi.

GOVINDA RAJAN, M. (1997): *Global Environmental Politics: India and North–South Politics of Global Environmental Issues*, Oxford University Press, Delhi.

GUNNAR, S. E. and J. EMMANUEL (1991): 'Choosing Policy Instruments for Pollution Control: A Review', Working Paper, World Bank, New York.

GUPTA, D. B., M. N. MURTY and R. PANDEY (1989): *Water Conservation and Pollution Abatement in Indian Industry: A Study of Water Tariff,* National Institute of Public Finance and Policy, Delhi.

HAMILTON, J. T. (1993): 'Politics and Social Costs: Estimating the Impact of Collective Action on Hazardous Waste Facilites', *Rand Journal of Economics*, vol. 24, no. 1.

HANEMANN, W. M. (1991): 'Willingness to Pay and Willingness to Accept: How Much Can They Differ?', *American Economic Review*, vol. 81, no. 3, pp 635–47.

——— (1994): 'Valuing the Environment through Contingent Valuation', *Journal of Economic Perspectives*, vol. 9, no. 4, pp 19–43.

HARDIN, G. (1968): 'The Tragedy of the Commons', *Science*, vol. 162, pp 1243–8.

HARRISON, D. (1983): 'The Regulation of Aircraft Noise', *in* T. C. Schelling (ed.), *Incentives for Environmental Protection*, M. I. T. Press, Cambridge, Mass.

HEYES, A. G. (1977): 'Environmental Regulation by Private Contest', *Journal of Public Economics,* vol. 63, pp 407–28.

——— (1994): 'The Economics of Strict and Fault Based Environmental Liability', *European Environmental Law Review*, vol. 3, no. 10.

HICKS, J. R (1946): *Value and Capital,* Oxford University Press, Oxford.

HOEL, M. (1990): 'Efficient International Agreements for Reducing Emissions of $CO_2$', Department of Economics, University of Oslo, mimeo. No. 6.

——— (1991): 'Global Environmental Problems: The Effects of Unilateral Actions Taken by One Country', *Journal of Environmental Economics and Management*, vol. 20, pp 55–70.

——— (1992): 'Carbon Taxes: An International Tax or Harmonized Domestic Taxes?', *European Economic Review*, vol. 36, pp 400–6.

——— (1997): '$CO_2$ and the Greenhouse Effect: A Game-Theoretic Exploration', *in* Dasgupta and Maler (eds), *The Environment and Emerging Development Issues,* Cambridge University Press, Cambridge.

JAMES, A. J. and M. N. MURTY (1996): 'Water Pollution Abatement: A Taxes-Standards Approach for Indian Industry', Working Paper No. E/177/96, Institute of Economic Growth, Delhi.

JANSEN, H. M. A. (1991): *West European Experiences with Environmental Funds*, Institute for Environmental Studies, Vrije Universiteit, Amsterdam.

JHA, R., M. N. MURTY and S. PAUL (1990): *Pricing of Postal Services in India*, National Institute of Public Finance and Policy, Delhi.

KAHNEMAN, D. and A. TVERSKY (1979): 'Prospect Theory: An Analysis of Decisions under Risk', *Econometrica*, vol. 47, pp 263–91.

KAHNEMAN, D. L., J. KNETSCH and R. THALER (1990): 'Experimental Tests of the Endowment Effect and Coase Theorem', *Journal of Political Economy*, vol. 98, pp 1325–48.

KNEESE, A. and B. BOWER (1968): *Managing Water Quality: Economics, Technology, Institutions,* The Johns Hopkins Press, Baltimore.

KNETSCH, J. L. (1994): 'Asking the Right Question: The Reference Point and Measures of Welfare Change', IEAS International Conference, Economic Perspectives of Pollution Control in the Pacific Rim Countries, Taipei, Taiwan, March.

KNETSCH, J. L. and J. SINDEN (1984): 'Willingness to Pay and Compensation Demanded: Experiemental Evidence of an Unexpected Disparity in Measures of Value', *Quarterly Journal of Economics,* vol. 99, pp 507–21.

KNETSCH, J. L. (1989): 'The Endowment Effect and Evidence of Non-reversible Indifference Curves', *The American Economic Review*, vol. 79, pp 1277–84.

———— (1994): 'Asking the Right Question: The Reference Point and Measures of Welfare Change', IEAS, International Conference, Economic Perspectives of Pollution Control in the Pacific Rim Countries, Taipei, Taiwan, March.

KNOESTER, A. and J. VAN SINDERN (1995): 'Taxation and the Abuse of Environmental Policies', *in* Boero, and Silberston.

KROELLER, E. (1993): 'Report on the Development Cooperation Directorate Development Centre Workshop on the use of Economic Instruments for Environmental Mnagement in Developing Countries', in *Economic Instruments for Environmental Management in Developing Countries*, Organization for Economic Co-operation and Development (OECD), Paris.

KRUTILLA, J. V. (1967): 'Conservation Reconsidered', *American Economic Review*, Sept. 1967, vol. 57, pp 777–86.

KRUTILLA, J. V. and A. C. FISHER (1975): *The Economics of Natural Environments*, Resources for the Future. The Johns Hopkins Press, Baltimore.

MARKANDYA, A. (1992): 'The Value of the Environment: A State of the Art Survey', *in* A. Markandya and J. Richardson (eds), *The Earthscan Reader in Environmental Economics*, Earthscan, London.

MATSUOKA, M. and B. ROSE (1994): *The DIR Guide to Japanese Economic Statistics*, Oxford University Press, Oxford.

MAY, P. H. and R. S. DA MOTTA (1996): *Pricing the Planet: Economic Analysis of Sustainable Development*, Columbia University Press, New York.

MEHTA, S., S. MUNDLE and U. SANKAR (1993): *Incentives and Regulations for Pollution Abatement with an Application to Waste Water Treatment*, National Institute of Public Finance and Policy, Delhi.

MILGROM, P. and J. ROBERTS (1988): 'An Economic Approach to Influence Activities in Organizations', *American Journal of Sociology*, vol. 94, Supp., S154–S79.

MILLER, J. (1988): 'Private Enforcement of Federal Pollution Control Laws', *Environmental Law Reporter*, vol. 14.

Ministry of Environment and Forests, Government of India (1974): *Water (Prevention and Control of Pollution) Act*, Delhi.

MISRA, S. (1996): 'Accounting for Costs of Water Pollution Abatement: A Case Study of Nandesari Industrial Area', Working Paper E/179/96, Institute of Economic Growth, Delhi.

———— (1997): 'Social Benefit Cost Analysis of Collective Action in Industrial Water Pollution Abatement: A Case Study of Nandesari Industrial Estate', mimeo. Institute of Economic Growth, Delhi.

MISRA, S. (1997): 'Measuring Benefits from Industrial Water Pollution Abatement: Use of Contingent Valuation in the Nandesari Industrial Area', Working Paper No. E/185/97, Institute of Economic Growth, Delhi.

MISRA, S. and M. N. MURTY (1995a): 'Collective Action in the Industrial Water Pollution Abatement: Conceptual Issues and Empirical Analysis', Working Paper No. E/166/95, Institute of Economic Growth, Delhi.

———— (1995b): 'Water Pollution Abatement by Isolated Small-scale Industries: Experiences of Industries in Sonepat Area of Haryana', mimeo., Institute of Economic Growth, Delhi.

MITCHELL, R. C. and R. T. CARSON (1989): *Using Surveys to Value Public Goods: The Contingent Valuation Method*, Resources for the Future, Washington, D.C.

MUKHERJEE, D. and I. PNG (1994): 'Corruptible Law Enforcers: How Should They be Compensated?', Discussion Paper No. 94–07, Indian Statistical Institute, Delhi.

MURTY, M. N. (1994a): 'Management of Common Property Resources: Limits to Voluntary Collective Action', Environmental and Resource Economics, *Journal of European Association of Environmental and Resource Economists*, vol. 4, pp 581–94.

———— (1994b): 'Optimal Control of Water Pollution: Pigouvian Tax Versus Coasian Free Market', IEAS International Conference, Economic Perspectives for Pollution Controls in the Pacific Rim Countries, 18–19 March.

———— (1995a): 'Environmental Regulaton in the Developing World: The Case of India', *Review of European Community and International Environmental Law*, Dec.

———— (1995b): 'Collective Action: A Deterrent to Colluding Industry and Corrupt Bureaucracy in the Water Pollution Abatement', Working Paper Series, No. E/165/95, Institute of Economic Growth, Delhi.

———— (1995c): 'Use of Economic Instruments for Controlling Pollution in India', Working Paper No. E/169/95, Institute of Economic Growth, Delhi.

———— (1996a): 'Fiscal Federalism Approach for Controlling Global Environmental Pollution', Environmental and Resource Economics, *Journal of European Association of Environmental and Resource Economists*, vol. 8, pp 449–59.

———— (1996b): 'Natural Resource Accounting: Measuring Value Added by Environmental Resources with Illustrartionas from Indian Data', Working Paper No. E/178/96, Institute of Economic Growth, Delhi.

———— (1996c): 'Role of Government in Environmental Management: Is it Catalytic or Coercive?', mimeo., Institute of Economic Growth, Delhi.

——— (1997): *Integrating Environmental Considerations in to Economic Decision Making: Institutional Arrangements and Mechanisms at National Level in India*, ESCAP, Bangkok.

MURTY, M. N. and P. B. NAYAK (1982): 'Externality Abatement Technologies, Pigouvian Taxes and Property Rights', *Indian Economic Review*, vol. 17, pp 1–26.

——— (1993): 'A Normative Approach to Resource Transfers in a Federal State', *in* A. Bagchi and N. Stern, *Tax Policy and Planning in Developing Countries*, Oxford University Press, Delhi.

——— (1995): 'Collective Action in Industrial Water Pollution Abatement: Experiences of Some Industrial Estates in South India', mimeo. Institute of Economic Growth, Delhi.

MURTY, M. N. and U. R. PRASAD (1977): 'Emission Reduction and Influence of Local Communities in India', mimeo., Institute of Economic Growth, Delhi.

MURTY, M. N. and R. RAY (1990a): 'Optimal Taxation and Resource Transfers in a Federal State', Working Paper, No. 90/01, Department of Economics, Delhi School of Economics, Delhi.

——— (1990b): 'Fiscal Federalism and Optimal Commodity Taxation: A Comparison of Centralized and Decentralized Decisions', Working Paper No. E/147/90, Institute of Economic Growth, Delhi.

NANDY I., R. A. DARYAPURKAR and S. N. KAUL (1991): *Common Effluent Treatment Plant—An Overview*, National Environmental Engineering Research Institute (NEERI), Nagpur.

NATIONAL PRODUCTIVITY COUNCIL (1992): *Report on Design of a Common Effluent Treatment Plant*, Nov.

NAYSNERSKI, W. and T. TIETENBERG (1992): 'Private Enforcement of Federal Environmental Law', *Land Economics*, vol. 68.

NEERI (1991): *Report on Environmental Pollution Caused by Patancheru and Bollaram Industrial Estates in Nearby Villages of Medak District of Andhra Pradesh, India*, Nagpur.

——— (1992), *Comon Effluent Treatment Plant: State of the Art*, Nagpur.

NOAA PANEL, (1993): 'Report of the NOAA Panel on Contingent Valuation', *U.S. Federal Register*, vol. 58, no. 10, pp 4602–14, 15 Jan.

NOZICK, R. (1974): *Anarchy, State and Utopia*, Basic Books, New York.

OATES, W. (1985): 'The Environment and the Economy: Environmental Policy at the Cross Roads', *in* J. Quigley and D. Rubenfeld (eds), *American Domestic Priorities: An Economic Appraisal*, University of California Press, Berkeley, CA. pp 311–45.

OATES W. and W. J. BAUMOL (1975): 'The Instruments for Environmental Policy', *in* E. Mills (ed.), *Economic Analysis of Environmental Problems*, Columbia University Press, New York, pp 95–128.

OATES, W. E. (1990): 'Public Finance with Several Levels of Government: Theories and Reflections', Paper Presented at 46th Congress, International Institute of Public Finance, Brussels, 27–30, Aug.

——— (1991): 'Pollution Charges as a Source of Public Revenue', Resources for the Future, Discussion paper No. QE92–05, Washington, DC.

O'CONNOR, D. (1992): 'The Use of Economic Instruments in Environmental Management: The Experience of East Asia', *Economic Instruments for Environmental Management in Developing Countries*, Proceedings of a workshop held at OECD Headquarters, Paris, 8 Oct. OECD, Paris.

——— (1993): 'The Use of Economic Instruments in Environmental Management: The Experience of East Asia', in *Economic Instruments for Environmental Management in Developing Countries*, OECD, Paris.

——— (1994): 'Managing the Environment with Rapid Industrialization: Lessons from the East Asian Experience', Pre-publication draft monograph, Development Centre, Organisation for Economic Co-operation and Development, Paris.

——— (1996): 'International Experience with Market-based Instruments', Paper presented at the International Seminar on Market-based Instruments for Pollution Prevention, organized by the Confederation of Indian Industries, New Delhi, India; International Institute for Environment and Development, London, UK & the United Nations; Industrial Development Organization, Vienna, Austria in New Delhi, July.

ODAGIRI, H. (1994): *Growth through Competition, Competition through Growth: Strategic Management and the Economy in Japan,* Clarendon Press, Oxford.

ODUAM, E. P. (1971): *Fundamentals of Ecology,* 3rd edn, Saunders, Philadelphia.

OECD (1972): *Annex to Recommendation on Guiding Principles Concerning International Economic Aspect of Environmental Policies,* OECD, Paris.

——— (1972): *The Polluter Pays Principle: Definition, Analysis Implementation*, OECD, Paris.

OLSON, M. (1965): *The Logic of Collective Action,* Harvard University Press, Mass. Cambridge.

OSTROM, E. (1990): *Governing the Commons: The Evolution of Institutions for Collective Action.*

PANAYOTOU, T. (1993): 'Economic Incentives for Environmental Management in Developing Countries', *in Economic Instruments for Environmental Management in Developing Countries,* OECD, Paris.

PARGAL, S. and D. WHEELER (1996): 'Informal Regulaton of Industrial Pollution in Developing Countries: Evidence from Indonesia', *Journal of Political Economy*, vol. 104, no. 6, pp 1314–27.

PARGAL, S., M. MANI and M. HUQ (1997): 'Inspections and Emissions in India: Puzzling Survey Evidence about Industrial Pollution', Policy Research Working Paper, No. 1810, Development Research Group, World Bank.

PARIKH, J., V. SHARMA, U. GHOSH and M. PANDA (1993): 'Trade and Environment Linkages: A Case Study of India', UNCTAD, Geneva.

PEARCE, D. (1991): 'The Role of Carbon Taxes in Adjusting to Global Warming', *Economic Journal*, vol. 101, pp 938–48.

PEARCE, D. W. and R. K. TURNER (1990): *Economics of Natural Resource and Environment*, Harvester Wheatsheaf, New York.

PESKIN, H. (1989): 'A Proposed Environmental Accounting Framework', *in* Y. L. Ahmed, S. E. Serafy and E. Lutz, (eds), *Environmental Accounting for Sustainable Development*, The World Bank, Washington D. C.

PEZZEY, J. (1994): 'On the Political and Informational Economy of Distributing the Efficiency Gains from Market Mechanisms of Pollution Control', University College of London, Working Paper, Oct.

PIGOU, A. C. (1932): *The Economics of Welfare*, 4th edn, London.

PORTNEY, P. R. (1994): 'The Contingent Valuation Debate: Why Economists Should Care?', *Journal of Economic Perspectives*, vol. 8, no. 4, pp 3–17.

POTERBA, J. (1991): 'Designing a Carbon Tax', *in* Dornbusch and Poterba.

RAJAH, N. and S. SMITH (1993): 'Taxes, Tax Expenditures, and Environmental Regulation', *Oxford Review of Economic Policy*, vol. 9, no. 4 pp 41–65.

RANDALL, A. and J. R. STOLL (1980): 'Consumer Surplus in Commodity Space', *American Economic Review*, vol. 70, June 1980, pp 449–55.

REPETTO, R., W. MAGRATH, M. WELLS, C. BEER and F. ROSSINI (1989), *Wasting Assets: Natural Resources in the National Income Accounts*, WRI, Washington D. C.

ROBERTS, M. J. and M. SPENCE (1976): 'Effluent Charges and Licences Under Uncertainty', *Journal of Public Economics*, vol. v.

ROSE-ACKERMAN, S. (1978): *Corruption: A Study in Political Economy*, Academic Press, New York.

RUNGE, C. F. (1981): 'Common Property Externalities: Isolation, Assurance and Resource Depletion in a Traditional Grazing Context', *American Journal of Agricultural Economics*, vol. 63, pp 595–606.

———— (1984): 'Strategic Interdependence in Models of Property Rights', *American Journal of Agricultural Economics*, vol. 66, pp 807–13.

———— (1986): 'Common Property and Collective Action in Economic Development', *World Development*, vol. 14, no. 5, pp 623–35.

SANDMO, A. (1975): 'Optimal Taxation in the Presence of Externalities', *Swedish Journal of Economics*, vol. 77, pp 86–98.

SARANGI, S. and G. COHEN (1995): 'A Tale of Two Industrial Estates', *Economic and Political Weekly*, vol. xxx, no. 24, pp 1419–22.

SAXENA, N. C. (1993): 'Social Forestry and Panchayats', *Seminar*, vol. 406, June.

SCHELLING, T. (1991): 'International Burden Sharing and Co-operation: Prospects for Co-operative Approaches to Global Warming', *in* Dornbush and Poterba.

SCHELLING, T. C. (1992): 'Some Economics of Global Warming', *American Economic Review*, March, pp 1–14.

SCOTT S. and J. LAWLOR (1994): *Waste Water Services: Charging Industry the Capital Cost,* The Economic and Social Research Institute, Policy Research Series, Paper No. 22, Dec.

SEN, A. K. (1967): 'Isolation, Assurance and the Social Rate of Discount', *Quarterly Journal of Economics*, vol. 81, no. 1, pp 112–24.

SENECA, JOSEPH J. and MICHAEL K. TAUSSIG (1984): *Environmental Economics*, 3rd edn, Prentice-Hall, New Jersey.

SHARMA, A. H. (1995): 'Environmental Impact Assessment, Along the Effluent Channel from Baroda to Jambusar and the Confluence with Mahi Estuary and the Gulf of Cambay with Special Reference to Heavy Metals', Ph.D. thesis, M. S. University, Baroda.

SHAW, E. (1994): 'The Dilemma of Environmental Protection and Economic Development in Taiwan: A Historical Perspective', paper prepared for the IGFA Mini-Workshop on the Newly Industrialized Countries and Global Change Research, Taipei, Taiwan (ROC), 24 Jan.

SOLORZANI, R. et al. (1992): *Accounts Overdue: Natural Resources Depreciation in Costa Rica,* World Resource Institute, Washington D. C.

STARRETT. D. A. (1972): 'Fundamental Non-Convexities in the Theory of Externalities', *Journal of Economic Theory*, vol. 4.

STIGLER, G. J. (1989): 'Two Notes on the Coase Theorem', *Yale Law Journal*, vol. 99, pp 631–3.

SUGDEN, R. (1984): 'Reciprocity: The Supply of Public Goods Through Voluntary Contributions', *The Economic Journal*, vol. 94, pp 772–87.

Supreme Court, Government of India (1990): W.P.(c) 1056/90, *Indian Council for Enviro–Legal Action and ors.* vs. *Union of India and ors.*

——— Vellore Citizens Welfare Forum vs. Union of India and others.

TAUB, FRIEDA M. (1987): 'Indicators of Change in Natural and Human-impacted Ecosystems: Status', *in* Sidney Draggan, John J. Cohrssen and Richard E. Morrison (eds), *Preserving Ecological Systems: The Agenda for Long-term Research and Development*, Praeger, New York.

TIETENBERG, T. H. (1986): 'Uncommon Sense: The Program to Reform Pollution Control Policy', *in* L. W. Weiss and M. W. Klass (eds), *Regulatory Reform: What Actually Happened,* Little Brown and Co., Boston, pp 269–303.

——— (1991): 'Economic Instruments for Environmental Regulation', *Oxford Review of Economic Policy*, vol. 6, no. 1, pp 17–33.

——— (1991): 'Economic Instruments for Environmental Regulation', *Oxford Review of Economic Policy*, vol. 6, no. 1, pp 17–33.

UN (1992): The United Nations Conference on Environment and Development, Rio de Janeiro, Brazil.

——— (1993a): *Handbook of National Accounting: Integrated Environmental Accounting Studies and Methods*, Series F. No. 61, Department of Economic and Social Information and Policy Analysis, United Nations, New York.

——— (1993b): 'Integrated Environmental and Economic Accounting', Interim version (Sales No. E.93 XVII. 12), United Nations, New York.

WADE, R. (1987): 'The Management of Common Property Resources: Collective Action as an Alternative to Privatization or State Regulation', *Cambridge Journal of Economics,* vol. 11, pp 107–32.

WALLACE, D. (1995): *Environmental Policy and Industrial Innovation: Strategies in Europe, the US and Japan*, The Royal Institute of International Affairs, Energy and Environment Programme, Earthscan Publications Ltd., London.

WEISBROD, B. A. (1964): 'Collective Consumpton Services of Individual Consumption Goods', *Quarterly Journal of Economics,* vol. 78, no. 3, pp 471–77.

WILLIG, R. D. (1976): 'Consumer Surplus Without Apology', *American Economic Review*, vol. 66, no. 4, Sept.

WB (1992): *World Development Report*, World Bank, Washington D. C.

WRI (1991): *Accounts Overdue: Natural Resource Depreciation in Costa Rica*, World Resource Institute, Washington D. C.

# Index